# AUTODESK®
# REVIT® 2023
# ARCHITECTURE

## LICENSE, DISCLAIMER OF LIABILITY, AND LIMITED WARRANTY

# AUTODESK®
# REVIT® 2023
# ARCHITECTURE

**Munir M. Hamad**
*Approved Autodesk Instructor*

MERCURY LEARNING AND INFORMATION
Dulles, Virginia
Boston, Massachusetts
New Delhi

Publisher: David Pallai
MERCURY LEARNING AND INFORMATION
22841 Quicksilver Drive
Dulles, VA 20166
info@merclearning.com
www.merclearning.com
(800) 232-0223

Munir M. Hamad. *AUTODESK® REVIT® 2023 ARCHITECTURE.*
ISBN: 978-1-68392-844-7

REVIT®  is a trademark of Autodesk, Inc. Version 1.0, 2023

The publisher recognizes and respects all marks used by companies, manufacturers, and developers as a means to distinguish their products. All brand names and product names mentioned in this book are trademarks or service marks of their respective companies. Any omission or misuse (of any kind) of service marks or trademarks, etc. is not an attempt to infringe on the property of others.

Library of Congress Control Number: 2022935274
222324321    Printed on acid-free paper in the United States of America.

Our titles are available for adoption, license, or bulk purchase by institutions, corporations, etc. For additional information, please contact the Customer Service Dept. at (800) 232-0223 (toll free).

All of our titles are available in digital format at academiccoursewareware.com and other digital vendors. *Companion disc files for this title are available by contacting info@merclearning.com.* The sole obligation of MERCURY LEARNING AND INFORMATION  to the purchaser is to replace the disc, based on defective materials or faulty workmanship, but not based on the operation or functionality of the product.

# CONTENTS

# BOOK PURPOSE AND OBJECTIVES

You can divide this book into two halves. The first thirteen chapters cover the basics of Revit Architecture 2023. The second half, from Chapter 14 to Chapter 25, covers the intermediate and advanced features of the software. This book demonstrates in very simple step-by-step procedures how to create a building model from scratch using all the tools the software offers.

At the completion of this book, the reader will be able to:

- Identify the difference between CAD & BIM
- Identify the different parts of Revit interface
- Draw and modify in Revit
- Prepare a new project
- Add and manipulate walls
- Insert doors and windows
- Create and manipulate curtain walls
- Create floors and roofs
- Add components and ceilings
- Add stairs, ramps, and railings
- Create views
- Create dimensions, text, and legends
- Visualize a model
- Create sheets and print them
- Create phases, design options, and path of travel
- Create and control toposurfaces
- Create Rooms and Areas
- Tag and create details
- Create Links, Import, and Export

- Create Masses, customize walls, floors, roofs, and ceiling families
- Create families
- Customize doors, windows, and railings
- Deal with worksets

# PREFACE

Revit Architecture is considered by many to be the best BIM solution in the international market. Autodesk acquired the company in 2002 and the Revit user base is adding hundreds of thousands of users annually. If you are studying at a university or work for a design or construction company, and have decided to use Revit Architecture, you've made the right choice.

This is a comprehensive book about Autodesk Revit 2023 Architecture. It will not teach the reader architecture design, which is a pre-requisite for this book. It can be used as an ***instructor-led*** or ***teach-yourself*** text-and-project book to help master the basics of Revit 2023 Architecture.

- We prepared two sets of exercises, one for metric units (mm) and the other for imperial (inch, ft).
- At the end of each chapter, you will find "Chapter Review Questions" that will help you test yourself.
- There are fifty-seven exercises integrated throughout the book, to help the reader implement what has been learned.
- Starting with Revit 2018, Autodesk changed the naming method and combined all of the Revit modules into a single product. So, when you buy the software, instead of Architecture, Structure, and MEP being separate modules, now one product contains all of them. Accordingly, this book teaches the Architecture part of Revit 2023.

# IN THE COMPANION FILES

The companion files included with this book contain:

- A link to the *Revit 2023 Trial* version, which will last for 30 days starting from the day of installation. This version will help you solve all the exercises and workshops in this book. *The student trial version can be extended beyond the 30-day period.*
- Practice files and projects which will be your starting point to solve all exercises and workshops in the book.
- Copy the folder named "Practices and Projects" onto the hard drive of your computer. In the Project folder, you will find two folders; *the first is called "Metric" for metric units projects and the second one is called "Imperial" for imperial units projects.*
- Companion files are also available by writing to the publisher at *info@ merclearning.com.*

# EXERCISES

- We prepared two sets of exercises, one for metric units (mm) and the other for imperial (inch, ft).
- The companion files include both sets of exercises. Copy the suitable set to your computer before you start reading this book.
- In the exercises, you will find the metric dimension first, then the imperial dimension within parentheses.
- The numbers are not necessarily equal.

## CHAPTER 1

# INTRODUCTION TO REVIT *2023*

## This Chapter Contains

- CAD vs. BIM
- What is Revit?
- Revit interface
- Zooming in Revit
- Creating new files and opening existing files

## CAD VS. BIM

- CAD means **C**omputer **A**ided **D**rafting:
  - When CAD was introduced in the early 1980s, it was revolutionary.
  - In CAD we draft; we do not design or model.
  - A line in CAD software could mean anything! It could mean the outer edge of a wall, an electrical wire, or a hot water pipe.
  - In CAD, plans, sections, elevations, and details should be drawn in separate files. The collection of these files will be the design of your project.
  - The sheets are not interrelated by any means except in the minds of the engineers who designed them.
  - If one part of the project changes, the engineers must revisit all the related drawings and edit them manually.
  - In CAD, groups of engineers from different disciplines use primitive commands to share their designs.

- BIM means **B**uilding **I**nformation **M**odeling.
  - This is the new approach to designing buildings.
  - BIM is based on the concept of creating your model in 3D right away, using intelligent elements like walls, doors, windows, floors, ceilings, and roofs. The user will not use lines, arcs, or circles.
  - Each element holds inside it an enormous amount of ***information***. For example: a wall drawn in the Floor Plan holds inside it its height; hence, you see it in 2D as a plan, but when you visit the 3D view you will see it extending in 3D as well.
  - Some elements because of their nature have the need to have a host, like a door and window need a wall to host them, and a lighting appliance needs a ceiling to host it.
  - The BIM approach supports *Parametric*, which means intelligent elements understand their relationships with each other; when one element changes, the other elements connected to it will react accordingly.
  - The BIM approach supports sharing the model information across users. Hence if an update is made for one user, all the others will know about it.
  - All parts of the model are interrelated; a change in one part will reflect on all other parts. For example, moving one wall in the Floor Plan view will affect all other views; hence there is no need to go to each sheet and change it manually. Even the schedule views that hold the information of the quantities of the different parts of this wall will react accordingly.

## REVIT ARCHITECTURE

- Revit Architecture is the best BIM solution for architects in the international markets these days.
- Revit Architecture is based on the idea of creating your whole model in a *single file* containing all the architecture elements; even if your model is very large, it will be in a single file.
- The Revit Architecture file will contain all the views of the 3D model that the user will create; hence floor plans, ceiling plans, sections, elevations, and schedules are all produced automatically.
- Revit Architecture is equipped with sharing and updating capabilities, which is unparalleled in today's software technology.

- Revit Architecture, Revit Structure (for structural engineers) and Revit MEP (for Mechanical, Electrical, and Plumbing engineers) are all in one package called Revit 2023; the three represent the ultimate BIM solution for all engineering disciplines.

## STARTING REVIT 2023

- Start Revit 2023 using the shortcut that appears on your desktop.
- You will see the following screen:

- In this screen you can create a new project or open an existing project.
- You can also open an existing Family file or create a new one.
- At the top left of the screen, you can see the Home button:

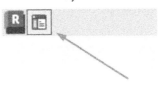

- You can use the Home button to switch between the home opening screen and the drawing area
- Once you create a new project or open an existing one, you will see the following:

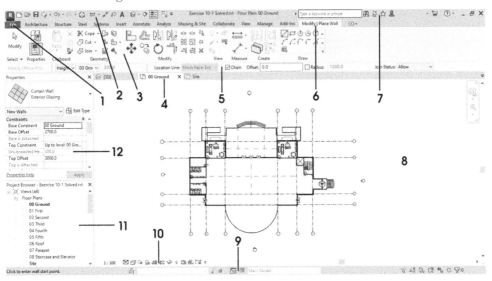

- The contents of this screen are as follows:

1. File Menu
2. Quick Access Toolbar
3. Ribbon
4. View tabs
5. Options Bar
6. Context Tab
7. Info Center
8. View Area
9. Status Bar
10. View Control Bar
11. Project Browser
12. Properties

## UNDERSTANDING THE REVIT 2023 INTERFACE

- The Ribbons and File Menu will be your primary methods to reach commands in Revit 2023.

### File Menu

- Click on the **File menu** and you will receive the following:

- You will find all file functions needed to:
  - Create a new project
  - Open an existing project
  - Save the current project
  - Save As the current project under a new name and/or a different folder
  - Export the current project to a different file format
  - Print the current file
  - Close all views of the current project, hence closing the project
  - Start Options dialog box
  - Exit Revit 2023

### Quick Access Toolbar

- This is the small toolbar hanging at the top left part of the screen:

- Using this toolbar, you can:
  - Open an existing file
  - Save the current file
  - Undo & Redo
  - Print
  - Create a PDF from Sheets and Views
  - Activate Controls and Dimensions
  - Measure between two references
  - Align Dimension
  - Tag by Category
  - Add Text
  - Open 3D View
  - Create a Section
  - Thin Lines
  - Close Inactive Views
  - Switch Windows
- If you click the arrow at the end, you will see the following:

- Using this list, you can add or remove commands. Also, you can **Customize the Quick Access Toolbar.**

**Ribbons**

- Ribbons consist of two parts:
  - Tabs
  - Panels

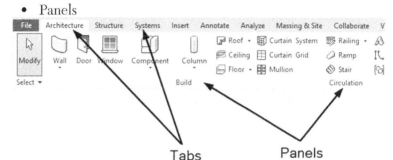

Tabs    Panels

- For instance, the Architecture tab consists of seven panels: Build, Circulation, Model, Room & Area, Opening, Datum, and Work Plane.
- Some panels will have a small triangle near the title to indicate additional buttons; if you click it then you will receive the following:

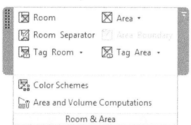

- Some buttons may have a small triangle at the right part, meaning there are more options, such as in the following illustration:

- If you stay for one second over any button, you will see a small help screen appear, as shown in the following:

- However, if you stay for three seconds, you will see an extended help screen. Refer to the following illustration:

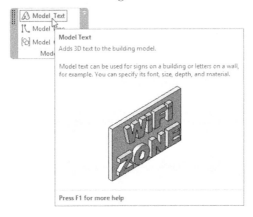

- Panels can be *docked* or *floating*. By default, all panels are docked. To make them floating, simply click the name of the panel, hold, and drag.
- While the panel is floating, there is a small button at the right side; the following image shows its function:

- Ribbons can be displayed differently on the screen by clicking the small arrow at the right repeatedly:
  - You can see the panels and their contents:

- It displays a single button with the name of the panel. If you hover over it, it will show the contents of each panel alone:

- It will show only the name of the panel. If you hover over it, it will show the contents of each panel alone:

- The name of the tabs will be displayed; once you click it, Revit Architecture will show all the panels related to that tab:

- If you click the arrow beside the small button, you will see:

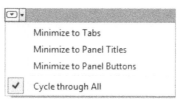

- In this list, you can pick any status you wish immediately without cycling.

### Info Center

- At the top right part of the screen, you will see the Info Center, as shown in the following:

- Info Center is the place where you can type any keyword(s) and Revit will search all its resources (online and off-line) to find you all the help topics related to your keyword(s).
- Other buttons are for signing into the Autodesk account, the Autodesk App Store, and the Help

### View Area

- Revit is based on views.

- When you start a new project based on one of the given templates (architectural as an example), you will find (at least) two floorplan views, two ceiling plans, and four elevations views.
- You will always work on a view, which is your place to input Revit elements.
- The first step to start creating your model is to ask yourself, "am I in the right view?" If yes, go ahead. If not, go to the Project Browser, double-click the desired view, and then start working.
- You can open as many views in your project file as desired.
- Each view will open in a view tab as shown below:

- This may slow down your computer; hence, you need to close views frequently. To close views use the (x) at the right of view name in View tab
- Or, you can go to Quick Access Toolbar and click the **Close Inactive Windows** button:

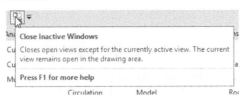

- Another way is to go to **View** tab and locate **Windows** panel, then click the **Tile** button to tile all the opened views:

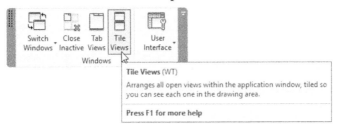

- To browse between the opened views, you can use one of the following:
  - [Ctrl] + [Tab]

- Go to the **View** tab, select the **Windows** panel, and click the **Switch Windows** button to see the following:

**NOTE** *If you have a multi-screen setup, you can click and hold the view tab, and move it to the second screen.*

*To move back to Tab Views, go to **View** tab, locate **Windows**, click **Tab Views**, as shown below:*

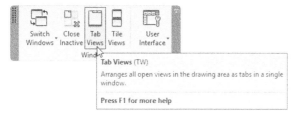

### Properties

- At the top left of the View Area, you will find **Properties** palette (of course, you can dock it anywhere else).
- This palette will show the properties of the selected element (Wall, Door, Roof, etc.). If there is no element selected, it will show the properties of the current view.
- Check the following two examples. The first one shows the properties of a wall, and the second one shows the properties of the current floorplan:

**Project Browser**

- Everything related to your project will be found in the Project Browser.
- You will find the following:
  - All views (floor plans, ceiling plans, sections, elevations, 3D, rendering, legends, etc.)
  - Schedules and Quantities
  - Sheets
  - Loaded families
  - Created groups
- The name of the current view will be bold compared to the other views.

**View Control Bar**

- This bar is located at the bottom of view area, above the status bar:

- In this bar, you can control the scale of the view:

- Use the Detail Level button to control the view; you have three choices:
  - Coarse
  - Medium
  - Fine

- Compare Coarse and Medium using the following figure (Wall):

Medium

Coarse

- The Visual Style button will show different styles for the 3D view (you can see the effect in 2D views as well but not as clear as the 3D):

- Other buttons will be discussed later, each in its proper place.

**Status Bar**

- The last bar at the bottom will show different things:
  - It will show you the next step after you start a command. For example, while you are in the Wall command, you will see at the left of the status bar "Click to enter wall start point."
  - The second part is the Workset part (advanced feature).
  - The third part is the Design Options part (advanced feature).
  - The rest are buttons for selecting elements and filters.

**Options Bar**

- When you start a new command or select an element, you will see this bar filled with information to help you finish your task correctly.
- For example, when you start the Wall command, the Options bar will look similar to the following:

Height ∨  00 Grot ∨  300.0        Location Line: Finish Face: Ext ∨  ☑ Chain  Offset: 0.0        ☐ Radius: 1000.0        Join Status: Allow        ∨

- These are all options related to creating a wall (which will be discussed later).

## ZOOMING IN REVIT

- The Mouse is the primary input device:
  - The Left Button is always used to Select or Click.
  - The Right Button has many purposes depending on what you are doing.
- The wheel in the mouse has zooming functions:
  - Zoom in by moving the wheel forward.
  - Zoom out by moving the wheel backward.
  - Pan by pressing the wheel and holding, then moving, the mouse.
  - Zoom to the extents of your model by double-clicking the wheel.
- You can use the Zoom commands available in the toolbar at the right (though we think that using the mouse for zooming is much easier). Click the second button and you will see the following menu:

**NOTE** *Zoom All to Fit works if multiple views are opened and tiled. It will Zoom all elements in all views in one shot.*

## 3D AND CAMERA IN REVIT

- When you create a new file using one of the premade templates, there will be no 3D views.
- When you issue the 3D command for the first time, a new view called {3D} will be created. The default 3D view is Orthographic
- To issue the 3D view command, go to the Quick Access Toolbar and click the arrow beside the **Default 3D View** to see the following list:

- Click the **Default 3D View** button to see the 3D presentation of your model.
- At the top right of your screen, you will see the ViewCube:

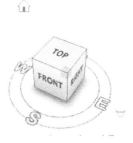

- The ViewCube will allow you to view the 3D model from six orthographical views and from the edges and corners of the cube.
- To orbit the 3D model, hold [Shift], click and hold the mouse wheel, and move the mouse.
- To create an additional interior or exterior 3D view, create a camera.
- To create a camera, go to Quick Access Toolbar, click the arrow beside **Default 3D View** and select **Camera** option:

- Specify two points:
  - Camera point
  - Target point
- Once done, Revit 2023 will create a new 3D view (with a temporary name) and will make it the current view.

**NOTE** *If you want to view your 3D model as* **Perspective***, right-click the ViewCube and select Perspective, or use Camera (all Camera views are Perspective).*

## STEERING WHEEL

- The Steering wheel will allow you to navigate the model using different methods like zooming, orbiting, panning, walking, etc.
- To issue the command, go to the toolbar at the right and click **Steering Wheel** button, as shown in the following:

- Choose **Full Navigation** option and you will see the following:

- There are eight viewing commands, four located in the outer circle, and four located in the inner circle. The commands in the outer circle are:
  - Zoom command
  - Orbit command
  - Pan command
  - Rewind command
- In the inner circle, you will find:
  - Center command
  - Look command
  - Up/Down command
  - Walk command

### Zoom

- This command will zoom in on the 3D (although using the mouse wheel will do almost the same job). Move your cursor to the desired location of your model, start the Zoom option, click and hold, move your mouse forward to zoom in, and move your mouse backward to zoom out.

### Orbit

- This command is similar to the Zoom command. Move your cursor to the desired location, select the Orbit option, click and hold and move the mouse right and left and up and down to orbit your model.

### Pan

- This command will Pan in 3D. Move your cursor to the desired location, select the Pan option, and click and hold and move the mouse right and left and up and down to pan over your model.

### Rewind

- This command will record all your actions that took place using the other seven commands. Use it as if you are rewinding a movie. Start the Rewind command and you will see a series of screen shots; click and drag your mouse backward to view all your actions.

### Center

- This command will specify a new center point for the screen. (You should always hover over an object.) Select the Center option, click and hold, and then locate a new center point for the current view; when done, release the mouse, and the whole screen will move to capture the new center point.

### Look

- Assume you are in a place and you are not moving; your head is moving up and down, left and right. Move the cursor to the desired location, select the Look option, click and hold, and then move the mouse right and left and up and down.

### Up/Down

- This command will go up above the model or down below the model. Move your cursor to the desired location and select the Up/Down option and you will see a vertical scale; click and hold and move the mouse up and down.

### Walk

- For this command to be active, you need to be in Perspective view. It will simulate walking around a place; you will have eight directions to walk through. If you combine this command with the Up/Down command, you will get best results. Move your cursor to the desired location, select the Walk option, and click and hold and move the mouse in one of the eight directions.
- The other versions of the Steering Wheel are mini copies of the Full Navigation:
  - The Mini Full Navigation Wheel shows a small circle that contains all eight commands.
  - The Mini View Object Wheel shows a small circle that contains the four commands of the outer circle in the full wheel.
  - The Mini Tour Building Wheel shows a circle that contains the four commands of the inner circle in the full wheel.
  - Basic View Object Wheel
  - Basic Tour Building Wheel

---

**NOTE**

- *While you are in the Perspective mode, you can use the Fly tool*
- *Fly tool will enable you to walk through a model and look around*
- *You can find it in the Navigation bar:*

---

## CREATE NEW PROJECT

- This command will create a new project based on a premade template.
- There are three ways to create a new project:
  - From the starting screen, click **New.**
  - From the Quick Access Toolbar, click **New** button.
  - From the File menu, click **New**, then **Project.**
- Using any of the above ways, you will see the following dialog box:

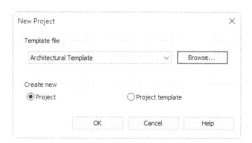

- To ensure that you are using the right template, click **Browse** button and you will see the following dialog box (there is another set of templates for imperial units):

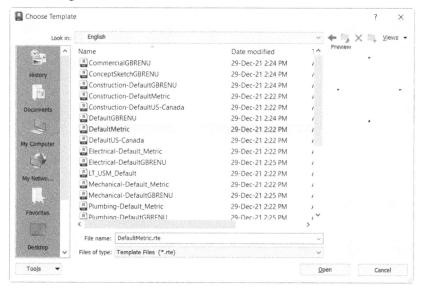

- Depending on the options you selected when you installed Revit, you will see the default templates or you will see other choices of templates.
- Revit template files have the extension *.rte*
- Select the desired template and click **Open** to start a new file.

<table>
<tr><td rowspan="4">**NOTE**</td><td>■</td><td>*Start a new architectural project using the architectural templates.*</td></tr>
<tr><td>■</td><td>*The first four templates are for architecture: the two that start with the word "Construction" and the two that start with the word "Default."*</td></tr>
</table>

## OPEN EXISTING PROJECT

- ▪ This command will open an existing project for further editing.
- ▪ There are three ways to open an existing project:
  - • From the starting screen, click **Open**.
  - • From the Quick Access Toolbar, click **Open** button.
  - • From File menu, click **Open**, then **Project**.
- ▪ Using any of these three ways, you will see the following dialog box:

- ▪ Specify the hard drive and the folder your project resides in. Revit project files have the extension **\*.rvt**.
- ▪ Click **Open** to open the desired project.

---

**NOTE** *While opening any file, this dialog box will tell you in which version of Revit this file was created in.*

---

## CLOSING FILE(S)

- ▪ To close all opened views of a project, hence closing the project itself.
- ▪ Using the File Menu, select **Close**.
- ▪ Select whether to save changes or not.

# EXERCISE 1-1    INTRODUCING REVIT 2023

**1.** Start Revit 2023.

**2.** Open the file **Exercise 1-1.rvt**.

**3.** Type the name of the current view: _____.

**4.** Double-click the view named 00 Ground. Using different zooming commands, explore this view.

**5.** Change the Detail level to Coarse, then Medium.

**6.** What is the scale of this view? _____.

**7.** Change it to 1:200 (1/16":1'), 1:50 (1/4":1'), then get it back to 1:100 (1/8":1').

**8.** Create a camera for different parts of the model.

**9.** Using the Quick Access Toolbar, click the 3D icon, then change the Visual style to Wireframe, Hidden, Consistent Colors, and then get it back to Shaded.

**10.** Go to the 01 First view and select one of the walls (by clicking it), avoiding curtain walls.

**11.** Check the Properties palette and record some of the information about this wall.

**12.** Using the Project Browser, double-click on the East Elevation view.

**13.** Using the View tab and the Window panel, tile all the opened views.

**14.** Use the Zoom All to Fit command and see its effect.

**15.** Select 00 Ground view. Click the Close Inactive Windows button to close all the other views.

**16.** Go to the 3D Section B-B and use the Steering Wheel to explore the 3D view. Change the view to Perspective so you can use both Walk and Fly

**17.** Using the File menu and Close command, close the current project without saving.

## NOTES

## CHAPTER REVIEW

**1.** One of the following is not Detail level:

    **a.** Coarse

    **b.** Fine

    **c.** Shaded

    **d.** Medium

**2.** BIM means _____

**3.** Project Browser contains all of the following except:

    **a.** Floor plans

    **b.** Schedules

    **c.** Scale of the current view

    **d.** Sheets

**4.** If there were no selected elements, the Properties palette will show nothing.

    **a.** True

    **b.** False

**5.** In CAD, we _____, we do not _____.

## CHAPTER REVIEW ANSWERS

**1.** c

**3.** c

**5.** Draft, Design

# 2

# HOW TO DRAW AND MODIFY IN REVIT

## This Chapter Contains

- Drawing commands and drawing techniques
- Selecting in Revit
- Some Modifying commands

## INTRODUCTION

- We will be exploring numerous drawing and modifying commands throughout the coming chapters.
- These commands share almost the same techniques, so we will dedicate this chapter to discussing these techniques only.

## DRAWING IN REVIT 2023

- ▪ While you are drawing walls (for example) in Revit 2023, the following will assist you in doing your job accurately:
  - Temporary dimensions
  - Alignment lines
  - Snaps
  - Draw context panel

### Temporary Dimensions

- ▪ Temporary dimensions appear in two different situations:
  - When you are drawing walls in floor plans, you will see the length of the wall and its angle measured from the east (180° counterclockwise, and 180° clockwise), as shown in the following example:

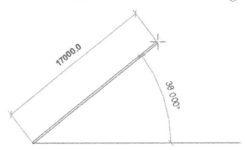

  - In this mode, you are allowed to input the value of the length by typing.
  - You will see the second type of temporary dimension after you finish drawing the wall. If you click the wall, you will be allowed to change the temporary dimension, as demonstrated in the following:

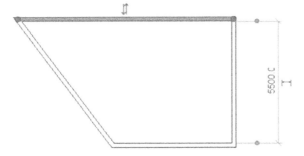

- Click the small blue circle to change the position of the dimension from the outer edge, inner edge, or center of the wall.
- The small shape beside the dimension is used to convert the temporary dimension to a permanent dimension:

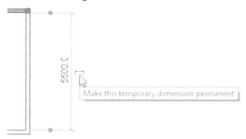

## Alignment Lines

- When you draw several walls, Revit will help you align the new wall to any existing wall, as shown in the following example:

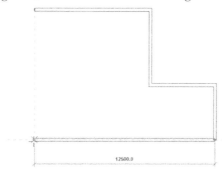

## Snaps

- Revit allows you to snap to any element, as in snapping to endpoints, the midpoint of a wall, or the intersection of two elements.
- By default all snaps are on.
- To control Snaps go to **Manage** tab, locate **Settings** panel, and click **Snaps** button:

■ You will receive the following dialog box:

■ In this dialog box you can do all/any of the following:
- Turn on/off any desired snap.
- Change the length dimension snap increment. The different values will work according to the zoom level. When you are closer you will use the lesser value.
- Change the angular dimension snap. The level of zoom applies here as well.
- Use the Temporary override by using the keyboard shortcuts; for instance, **SC** is used to snap to the center of an arc or circle element.

## Draw Context Panel

■ When you start any command in Revit that includes drawing, you will see receive the following (depending on the command, you may see less or more tools):

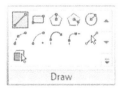

- The panel includes tools to help draw shapes such as:
  - Line tool, to draw straight shapes. If you want to draw continuous lines, make sure **Chain** option (which appears in the Options bar) is turned on.
  - Rectangle tool, to draw a rectangle shape using two opposite corners.
  - Two types of polygons (Inscribed and Circumscribed)
  - Circular shape
  - Three types of circular arcs
  - Fillet arc (to add an arc to two lines)
  - Select existing lines
  - Pick Faces
- More of these tools will be discussed when we tackle other commands.

## SELECTING IN REVIT

- There are three ways to select elements in Revit Architecture:
  - Click on the desired element.
  - Click on an empty space, hold the mouse, and go to the right to form a *Window*. Elements contained *fully* inside the Window will be selected.
  - Click on an empty space, hold the mouse, and go to the left to form a *Crossing*. Elements crossed by or contained fully inside the Crossing will be selected.
- The following is an example using the Window technique:

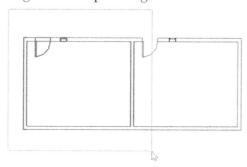

■ The following is an example using the Crossing technique:

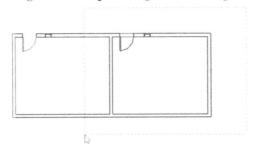

■ You can add and remove from the selection set as follows:
- Hold [Ctrl] key to add more elements to the selection set. You will see the cursor with the add sign.
- Hold [Shift] key to remove elements from the selection set. You will see the cursor with the subtract sign.
- Depending on what method you use, a green context tab will appear, which resembles the following:

■ Under the panel called **Modify**, you will see all modifying commands, like Move, Copy, Rotate, and so on. This method of selecting elements before issuing the modifying command is favorable when using Revit.
■ At the right under **Selection** panel you can see **Filter** button. This button is very important when you select multiple elements; it will help you segregate the contents as you wish. Click **Filter** and you will see the following dialog box:

- In the above example, your selection contains five walls, two doors, and two windows. You can use the two buttons at the right to either check them all or uncheck them all, or you can use the checkbox at the left to make your own decisions.

### Delete Command

- When you want to delete any element in Revit, simply select it and press the [Del] key on the keyboard.

## EXERCISE 2-1   DRAWING AND SELECTING

1. Start Revit 2023.

2. Open the file **Exercise 2-1.rvt**.

3. Change the Detail Level to Medium.

4. Start Wall command.

5. Make sure you are using Line tool from **Draw** context panel.

6. Somewhere between the East and South elevation symbols, click and go to the left, make sure you are using angle of 180°, and set the distance to 20000 (66').

7. Go up and set the distance to 16000 (52').

8. Go to the right and set the distance to 4000 (14').

9. Select **Start-End-Radius** Arc. The start point is already known, so go to the right to a distance of 12000 (38'), click to set the endpoint of the arc, and then move up to specify the radius which is 6000 (19') (the mouse will snap to the right distance).

**10.** Change the tool to Line and complete the following shape:

**11.** Click the right vertical wall to discover that the inside-to-inside distance of the walls is 19650 (64'-10 1/8"). Change this value to 21000 (66'-6").

**12.** Start Wall command, but this time, change the Type to Generic 225 Masonry (Generic - 8" Masonry).

**13.** Measuring from the lower horizontal wall from its left, click to start the wall at a distance of 3000 (9'-6"), then go to the right by 3000 (9'-6"), and finally close down to the horizontal wall to create the following shape:

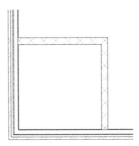

**14.** Make sure that the inside-to-inside distance of the room is 3000x3000 (9'-6" × 9'-6")

**15.** Start Door command and add a door like the following, making sure the distance between the wall and the door is 200 (1'):

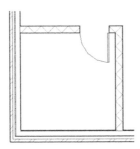

**16.** Add two windows like the following, bearing in mind that the two windows are exactly at the mid-distance between the walls:

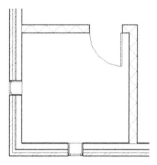

**17.** Using Window selection method, select only the interior walls, the door, and the two windows. Click **Filter** in the Multi-Select context tab, and deselect Walls and Doors. What are the selected elements now?

_____

**18.** Press [Esc] to release the selection. Double-click the wheel to Zoom All.

**19.** Save the project and close.

## SOME MODIFYING COMMANDS

- As we said previously, when you select element(s), the context **Modify** panel will appear, along with some element-specific functions.
- This method is the recommended method to modify elements in Revit.
- The context panel looks similar to the following:

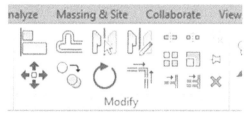

- We will discuss the following commands:
  - Move
  - Copy
  - Rotate
  - Mirror (two types)
  - Array (two types)

### Move Command

- Move command is always available when you select an element. For example, select a wall, then hover over it, and you will see the following:

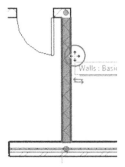

- Once you see this symbol, simply drag the element, and the temporary dimension will help you specify the exact distance.
- Another way is to select the desired element(s), then click **Move** button from **Modify** context panel:

- The selected element will appear similar to the following:

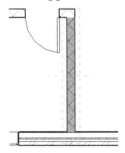

- The option bar will show the following:

Modify | Walls   ☐ Constrain  ☐ Disjoin  ☐ Multiple

- The two options are:
  - **Constraint:** To restrict the movement of the selected element to be either horizontal or vertical (this option is off by default).
  - **Disjoin:** To understand this option, note that the movement of walls is always stretching, which means the connected walls will always react by increasing the length or decreasing it. With **Disjoin** on, only the selected wall will move (this option is off by default).
- Specify the options, specify the start point, and then specify the end-point.

### Copy Command

- Copy command will copy the selected element(s).
- Everything that applies to the Move command applies to the Copy command, except for one; in the Options bar, there is an option called **Multiple** that allows multiple copies in the same command.
- Select the desired element(s), and then click **Copy** button from the Modify context panel:

### Rotate Command

- Use Rotate command to rotate element(s) around a specified point called **Center of rotation**. Select the desired element(s), and then click **Rotate** button from **Modify** context panel:

- At Options bar, you will see the following:

Modify | Walls ☐ Disjoin ☐ Copy Angle: Center of rotation: Place Default

- The Center of rotation is always the center of the selected element(s); to change it, click **Place** button and specify a new point:

Center of rotation

- Select **Disjoin** if you want the selected element(s) to disjoin from the attached elements.
- Select **Copy** if you want to keep the original and rotate a copy of it.
- You can specify an angle by typing, bearing in mind that positive values will rotate the elements counterclockwise.
- Another way to rotate is using the mouse; specify two angles (from the center of rotation) and Revit will calculate the net angle.

### Mirror Command

- The Mirror command will mirror the selected element(s) around the mirror line. There are two types of Mirror command:
  - **Mirror Pick Axis**, which uses an existing axis. Click **Mirror-Pick Axis** button from Modify context panel:

  - **Mirror Draw Axis**, which asks the user to specify two points as the mirror axis. Click **Mirror-Draw Axis** button from Modify context panel:

- The Options bar will appear as the following:

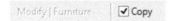

- By default, Revit will mirror and delete the original element. If **Copy** option is turned on, both (original and mirrored image) will stay.

## Array Command

- There are two types of Array in Revit 2023:
  - Linear Array
  - Radial Array
- Select the desired element(s), then click **Array** button from **Modify** context panel:

- In Options bar, you will see the following:

## Linear Array

- Linear Array is the default array type.
- It will repeat the selected items in linear fashion in any angle and distance desired by the user.
- Change one or all of the available options:
  - **Group and Associate** means all elements resulted from the array will be united in a single object. This means editing the whole array in the future is available.
  - **Number** means to specify the total number in the array including the original element.
  - **Move to 2nd or Last**, means the distance you will specify graphically will be from the original element to the 2nd, or to the last element of the array.
- After you specify the number of items and whether the distance will be measured to the 2nd or to the last element of the array, specify the distance and angle of the array.
- For example, refer to the following illustration:

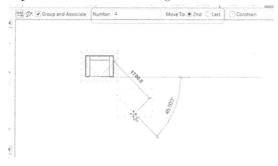

- In this example, the number is 4, the distance between the first and second is 1700, and with an angle of 45, the result will be similar to the following:

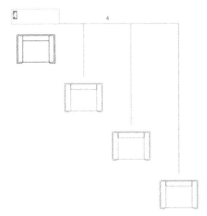

- You have the option to change the total number of items in the array. Input the new number and press [Enter].
- Even after the array command is finished, you can select any element of the array and you will see the following:

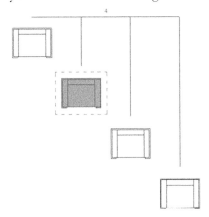

- While in this mode, you can change the total number of the array again, but you can also ungroup these elements; you will see a context tab named **Modify | Model Groups**, with one panel named **Group** containing two buttons as shown in the following:

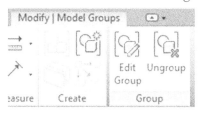

- Click the **Ungroup** button to ungroup elements from the group—the user should note that this would remove the array effect from the elements.

### Radial Array

- Radial array will repeat the selected element(s) in circular or semicircular fashion. After selecting **Radial** button, you will see the following in Options bar:

- By default, a Radial array needs a **Center of rotation** point (just like **Rotate** command); use **Place** button to specify a new **Center of rotation**.
- Specify whether to **Group and Associate** the total number of elements and whether the angle given (graphically) will be between the first and 2nd or last element.
- The user can type the angle (if you type the angle, the positive is counterclockwise) or specify it graphically.

## EXERCISE 2-2   SOME MODIFYING COMMANDS

**1.** Start Revit 2023.

**2.** Open the file **Exercise 2-2.rvt**.

**3.** Zoom to the room at the lower left corner.

**4.** Select the right wall and the door, and move them to the right by 500 (0'-11").

5. Select the two inside walls, the door, and the south window only. Start the Copy command, make sure that **Multiple** is on, specify the inside lower left corner as the first point, and then copy the room twice.

6. You should have the following:

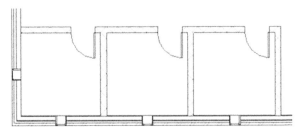

7. Using Mirror – Draw Axis command, mirror the three rooms (with their doors and windows) to the right side of the building.

8. For the middle room, delete one of the middle walls and correct the length of the horizontal wall to attach to the vertical wall.

9. Zoom to the middle of the building where the table and chairs are.

10. Rotate the left chair so it will be like the chair on the other side.

11. Array the top chair to be three chairs with the distance between chairs equal to 700 (1'-9").

12. Using Mirror – Pick Axis, mirror the three chairs to the other side, using the centerline of the table to be the mirror axis.

13. You should have the following image:

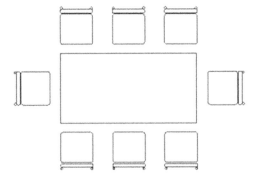

14. Zoom to the top of the building near the arc wall.

15. Select the only window there using the Array command with Radial type.

**16.** Change the Center of rotation to be the center of the arc wall.

**17.** Specify the number to be six. Specify the angle to be from the first to last element.

**18.** Pick the midpoint on the existing window, then go to the right measuring 150°.

**19.** Accept 6 as the final number. Select one of the windows, and change the total number to 8.

**20.** You should have the following:

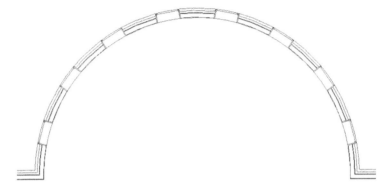

**21.** Save and close the file.

## NOTES

## CHAPTER REVIEW

1. A temporary dimension will be shown twice:

    **a.** No, only once, after the drawing is done.

    **b.** No, only once, while we are drawing.

    **c.** Yes, when we are drawing, and it appears automatically after drawing.

    **d.** Yes, when we are drawing, if you clicked the element after drawing.

2. If we do not turn Disjoin checkbox on, Move will be similar to _____

3. There are two types of Array and one type of Mirror in Revit Architecture:

    **a.** True

    **b.** False

4. The favorable method is to select elements first, and then issue the modify command.

    **a.** True

    **b.** False

5. One of the following statements is wrong:

    **a.** Array has two types: Rectangular and Radial.

    **b.** Mirror – Pick Axis is one of two types of Mirror command.

    **c.** Mirror – Draw Axis is one of two types of Mirror command.

    **d.** Array has two types: Linear and Radial.

6. _____ is the center of the element for both Rotate and Radial Array.

## CHAPTER REVIEW ANSWERS

**1.** d

**3.** b

**5.** a

CHAPTER 3

# *PROJECT PREPARATION*

## This Chapter Contains

- Project Information and Project Units
- Creating Levels
- Importing a CAD file
- Creating Gridlines
- Creating Columns

## INTRODUCTION

- The steps to prepare for a new project are:
  - Project Information and Project Units
  - Creating Levels
  - Importing a CAD file
  - Creating Gridlines
  - Creating Columns

# PROJECT INFORMATION AND PROJECT UNITS

- After you create a new file using the desired template, you may need to visit these two areas: Project Information and Project Units.
- Some companies may opt to include these two items in their template files.

### Project Information

- In Project Information dialog box, input general information about your project.
- This information will be used later in the final printed sheets.
- To issue this command, go to **Manage** tab, locate **Settings** panel, and then click **Project Information** button:

- The following dialog box will appear:

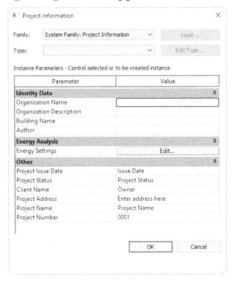

- The user should enter all or any of the following information:
  - Organization Name
  - Organization Description

- Building Name
- Author (the creator of the project)
- Energy Settings (to be filled if the project will be taken to Revit MEP)
- Project Issue Date
- Project Status
- Client Name
- Project Address, Name, and Number

## Project Units

- When you start a new project using a template file, this will dictate units to be used in the project. Most likely the default settings will be adequate.
- In this part, we will learn how to change the project units.
- To issue this command, go to **Manage** tab, locate **Settings** panel, and then click **Project Units** button:

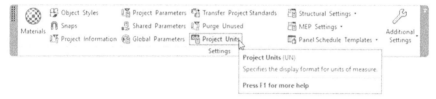

- You will see the following dialog box (the left for metric and the right for imperial):

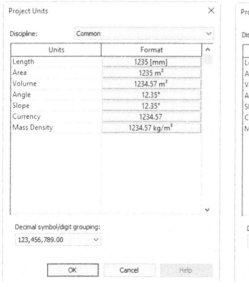

- You can change any of these settings by clicking Length button, which will produce the following dialog box:

- In the above dialog box, you can change all or any of the following:
  - Units
  - Rounding
  - Unit symbol

## CREATING LEVELS

- Creating levels is an important step to start a new project in Revit.
- All walls, columns, roofs, and ceilings will be constrained by levels.
- If you start with the default template, two levels separated by 4000mm (10') are available.
- Each level has three components:
  - Name
  - Height
  - View names associated with it
- You should be in the elevation or section view to create/edit a level.
- Change the name and height of any existing level.
- You can add any number of new levels.
- To create a new level, go **Architecture** tab, locate **Datum** panel, and click **Level** button:

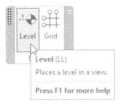

- ■ The following will change:
  - • You will see the following context tab, which includes two tools, drawing lines and picking existing lines:

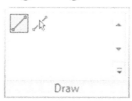

  - • The Options bar will show the following:

  - • Click on the **Make Plan View** checkbox to help you create a floor plan and ceiling plan for each newly added level.
  - • The Offset field will allow the user to set up an offset value for the Pick lines tool in the Draw context panel.
- ■ Ultimately, you can draw using the line tool. However, with the existence of the Pick line tool with the Offset value, drawing will be much easier. Since we have two existing levels, all you have to do is to click the **Pick line** tool, type a value in the Offset value, and then move your mouse to the desired level. You will see a preview dashed line above or below your existing level; click wherever you want it to be:

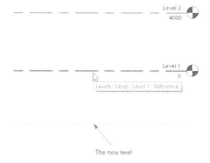

- ■ To change the name of a level, click the name twice and type in the new name. When done press [Enter] and you will see the following dialog box:

- Most likely your answer will be Yes, as the name of the views should be aligned with the name of the level. If you believe that this question is not important and you want to see it next time you do the same action, click the checkbox "Do not show me this message again" on.
- A few remarks about levels:
  - Levels with corresponding views will appear blue, levels without views will appear black.
  - By default level bubbles will move together using the small circle that appears to the left (or right) of the level bubble as shown below:

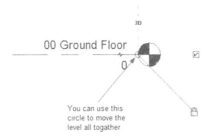

  - You can add an elbow to the level bubble using the **Add elbow** tool as shown below:

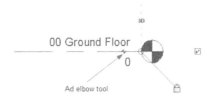

  - After adding an elbow, adjust all the parts by moving the small blue circles.
  - Show the bubble in both sides or in one side, using the checkbox at the left or right, as shown in the following:

- In addition, you can adjust the level family to show bubbles on both sides. Try the following:
  - Start Level command.
  - Using Properties palette, click **Edit Type** button.

- You will see the following dialog box:

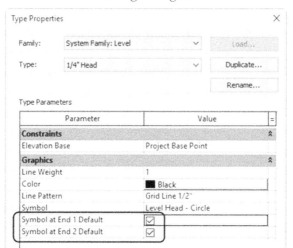

- Make sure that both Symbol at End 1 Default and Symbol at End 2 Default are checked.

**NOTE** *When you create a new level, click Plan View Types button on the Options bar, as shown below:*

- You will see the following dialog box; deselect the Structural Plan option so only Floor Plans and Ceiling Plans will be created:

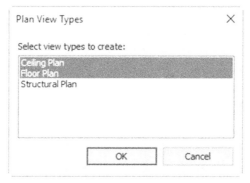

**NOTE**
- *By default, 3D view will contain 3D levels.*
- *All level functions in 2D are applicable in 3D as well*

# EXERCISE 3-1   PROJECT INFORMATION, UNITS, AND ADDING LEVELS

1. Start Revit 2023.

2. Start a new file based on DefaultMetric.rte template (default.rte template).

3. Save the file under the name Exercise 3-1_Solved.rvt

4. Go to **Manage** tab and click **Project Information** button to input the following data:

    a. Organization Name = XYZ Design

    b. Building Name = ABC Tower

    c. Author = Type your name

    d. Project Status = Design

    e. Client Name = ABC Invest

    f. Project Address = 12345 Boca Raton, FL, USA

    g. Project Name = ABC Tower

    h. Project Number = A-0909

5. Go to Project Units, take a look but do not change anything, and then click OK.

6. Go to South elevation view.

7. Do the following (answering Yes to renaming any view):

    a. Change the name of Level 1 to *00 Ground*.

    b. Change the name of Level 2 to *01 First*.

    c. Change the height of 01 First to 5500 (18').

    d. (Make sure you are not creating structural plans). Add new levels 3600 (12') apart with the names: *02 Second*, *03 Third*, *04 Fourth*, *05 Fifth*, and *06 Roof*.

    e. Add a new level over 06 Roof with 1000 (3'-6") in height and call it *07 Parapet* <u>without</u> creating a plan view for it.

**f.** Add a new level over 06 Roof with 3600 (12') in height and call it **08 Staircase and Elevator** without creating a plan view for it.

**g.** Using the Add elbow tool to create the following:

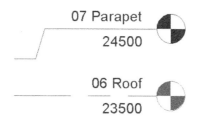

07 Parapet
24500

06 Roof
23500

**8.** Select one of the levels you just created. Click the **Edit Type** button and turn on both the Symbol at End 1 and 2 Defaults.

**9.** Go to 3D view to view 3D levels

**10.** Save and close the file.

## IMPORTING A CAD FILE

- This is an optional step, yet if you want to use it, it will be very useful in speeding up your creation process.
- This option will help you setup the grids, columns, and walls quickly using a CAD file.
- The advantage of this step is utilizing the huge experience of CAD (in general and AutoCAD in particular) software in the design offices.
- To import a CAD (or AutoCAD) file do the following:
  - Go to the desired floor plan.
  - Go to **Insert** tab, locate **Import** panel, and click **Import CAD** button:

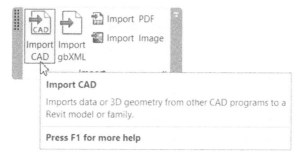

- The following dialog box will appear; go to the desired folder containing your file and select it (without pressing [Enter]):

- Select whether the CAD file will be used in the **Current view only**.
- Select the **Colors** of the file: Preserve, Black and White, or Invert.
- Select to import which Layers: All, Visible, or Specify.
- Select **Import Units** (leave it on Auto-Detect, as Revit can bring in the right units of the CAD file with high accuracy).
- Select Positioning: Automatic (Center-to-Center or Origin-to-Internal Origin).
- Select to Place the CAD file in which level (you have to be at the higher level in order to place it in lower levels)
- Click Open to finalize the process.
- The import will be one object. When you hover your mouse over it, a dark blue frame will appear. You will see the following:

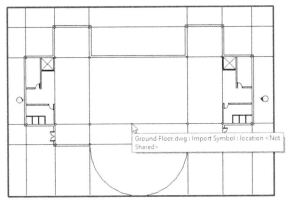

- We will use this later as tracing paper for our Revit elements; hence, we may need to obtain a more transparent look for the import file. To do that, right-click the import file, select the **Override Graphics in View** option, then select **By Element** or **By Category** (if you select By Element, only this instance will be affected; using By Category means all instances of this element in this view will be affected):

- You will see the following dialog box; pick **Halftone** to make it less dark:

**NOTE** *CAD Import command is not only for AutoCAD files, you can import several other files, as the below image suggests:*

- The file list includes: DWG, DXF, AXM, SAT, DGN, OBJ, 3DM, SKP, and STL.

## CREATING GRIDLINES

- Adding gridlines will help the user early on to set up the columns, which will be a guide for wall location as well.
- You can add gridlines using one of two ways:
  - By drawing
  - By picking existing lines coming from a CAD import

- To add a new gridline use the following method:
  - Go to the desired floor plan.
  - Go to **Architecture** tab, locate **Datum** panel, and then click **Grid** button:

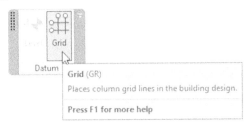

  - In context tab, select one of the available drawing tools as shown below:

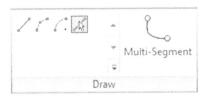

  - In Options bar, you will see the following:

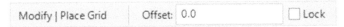

  - You can set the Offset value (both for drawing and picking lines), and you can choose to lock the gridline to the existing lines.
  - The first line drawn or picked will be numbered as 1, which you can change to any other value (a, or A, etc.); the next gridline will continue counting from that value.
- A few remarks about gridlines:
  - By default gridline heads will move together using the small circle below the bubble.
  - You can add an elbow to the gridline bubble using the Add elbow tool.
  - After adding an elbow you can adjust it by moving the small blue circles.
  - You can show the bubble on both sides or on one side.

NOTE
- *By default, 3D view will contain 3D Grids.*
- *All level functions in 2D are applicable in 3D as well.*

## CREATING COLUMNS

- There are two types of columns in Revit:
  - Structural Column
  - Architecture Column
- The architecture column is used as decorative column, whereas the Structural column is a standard column and can be used later by the structural engineer as a starting point for the structural system of the building.
- To input a structural column, do the following:
  - Go to **Architecture** tab, locate **Build** panel, click **Column** pop-up list, and then select **Structural Column** button:

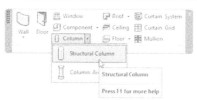

  - From Properties select the desired family:

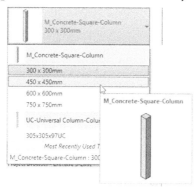

  - You will see the following context tab:

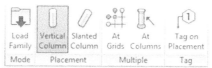

  - Under **Placement**, make sure you are using **Vertical Column** button.

- Under **Multiple** (and if you have already placed the gridlines), click **At Grids** button.
- If you want to add tags once you add columns, click **Tag on Placement** button.
- If you select **At Grids**, you will see the following context tab:

- **At Grids** will ask you to select the desired gridlines. Once you are done selecting, click **Finish** button to end the process of inputting columns.
- At the Options bar you will see the following:

- Select Height (to go up from the current level, choose **Height**, to go down from the current level, choose **Depth**).
- Specify the level to which the column will extend.

## EXERCISE 3-2    IMPORT CAD, GRIDLINES, AND COLUMNS

**1.** Start Revit 2023.

**2.** Open the file **Exercise 3-2.rvt**.

**3.** Go to **00 Ground** view and import the AutoCAD file called Ground Floor.dwg, using the following information:

    **a.** Colors = Black and White

    **b.** Layers = All

    **c.** Import Units = Auto-Detect

    **d.** Positioning = Origin-to-Internal Origin

    **e.** In the current view only

**4.** Using the same settings, import in 01 First view the AutoCAD file called First Floor.dwg.

5. If needed move the four elevation symbols to the proper places.

6. Set the Graphics Override = Halftone for both CAD Imports.

7. Use the lines from the CAD import in the 00 Ground view to input grid-lines, setting the vertical gridlines to A, B, . . ., and the horizontal gridlines to 1, 2, . . . (make sure to show bubbles on both sides for all gridlines).

8. Create structural columns using all the gridlines starting from 00 Ground up to 06 Roof; the size to be used is M_Concrete-Square-Column 300x300 (Concrete-Square-Column 12 × 12).

9. Delete the unnecessary columns: A1, A4, F1, and F4.

10. Go to the South elevation view and adjust the length of the levels to go beyond the columns at the right and at the left.

11. Save and close the file.

# NOTES

## CHAPTER REVIEW

**1.** A template file will dictate the units used:

   **a.** True

   **b.** True, but you can change them

   **c.** False

   **d.** False, template files do not contain any units

**2.** There are _____ types of columns in Revit Architecture.

**3.** Using levels:

   **a.** A template file includes two levels.

   **b.** You can change the name of the level.

   **c.** You can change the height of the level.

   **d.** All of the above.

**4.** While importing a CAD file, you should take all of the layers coming with the CAD file:

   **a.** True

   **b.** False

**5.** Project Information will appear in the sheets later on:

   **a.** True

   **b.** False

**6.** To make both levels and grid bubbles look good and not cover each other, you can use the _____ tool.

## CHAPTER REVIEW ANSWERS

**1.** b

**3.** d

**5.** a

## CHAPTER 4

# ALL YOU NEED TO KNOW ABOUT WALLS

**This Chapter Contains**

- Wall families
- Inputting walls
- Wall controls and temporary dimensions
- Additional modifying commands
- Wall profiles
- Hide and Unhide elements
- Slanted Walls
- Tapered Walls

## INTRODUCTION

- In this chapter, we will cover almost all aspects and features of adding walls in Revit Architecture.
- We will cover how to input walls in a Revit file and will discuss what features you need to control.
- Then we will cover the rest of the Modifying commands, which we started in Chapter 2.
- Finally, we will conclude by discussing wall profiles, Slanted Walls, and Tapered Walls

## WALL FAMILIES

- There are two types of families in Revit:
  - System families
  - Component families

### System Families

- They are not loadable in the current file.
- There is not an RFA for each family.
- You can create TYPES of system families within your project.
- System families are families that exist in a Revit project file as an essential part of your modeling process, such as walls, floors, ceilings, roofs, stairs, and railings.

### Component Families

- They are loadable in the current file.
- There is an RFA for each family.
- You can create new families using the Family Editor.
- Examples of Component families are: Furniture, Lighting Fixtures, Doors, Windows, Planting, Plumbing Fixtures, and so on.
- Wall families are System families and exist in all Revit project files by default.
- There are three types of wall families:
  - Basic Wall
  - Curtain Wall
  - Stacked Wall
- It is important to reiterate the concept that BIM elements are not like CAD objects. Walls in Revit are not only two parallel lines, but:
  - They have 2D presentation and 3D presentation.
  - They carry data (information) like width and height.
  - They consist of layers of materials (each layer has a width and function).
  - They can host other elements like Doors, Windows, and Light Fixtures.
  - They understand the relationship between elements and the roofs, floors, and ceiling attached to them.
  - You can generate a schedule from them.

_____

**NOTE** *To see the layers of the wall, change the Detail level to Medium or Fine.*

## INPUTTING WALLS

- There are two ways to input walls in Revit, either by using the drawing tools, or—if you have a CAD Import—by picking existing lines from a CAD file.
- To input walls, go to **Architecture** tab, locate **Build** panel, select **Wall** button, and then select **Wall: Architecture**:

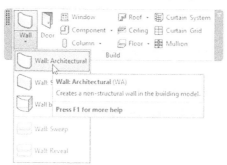

- The properties will show the wall types so you can select one of them:

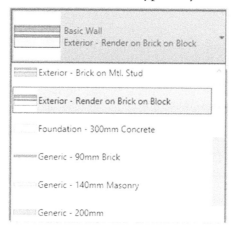

- You will see the following context tab:

Draw

- Options bar will show the following:

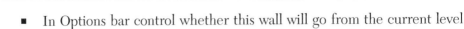

- In Options bar control whether this wall will go from the current level downward (choose **Depth**) or upward (choose **Height**).
- You have two options to control the height of the wall starting from current level:
  - Unconnected, which means the wall is not connected to a level. In this case input the height value.
  - Or to choose the level you want the wall to reach, refer to the following:

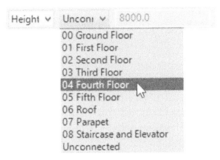

- To continue discussing the height concept, you can set the Base Constraint (the starting level) and Base Offset (which can be positive or negative), then the Top Constraint (the ending level) and Top Offset (which can be positive or negative):

■ To illustrate the concept of Base and Top Constraints along with positive and negative offsets, check the following (numbers are in mm):

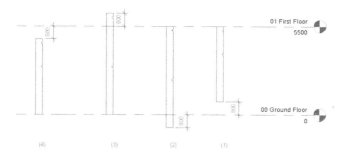

■ In the previous illustration, there are four cases:
- **Case (1):** the base offset is a positive offset
- **Case (2):** the base offset is a negative offset
- **Case (3):** the top offset is a positive offset
- **Case (4):** the top offset is a negative offset
■ Specify the Location Line:

■ To understand the above, in Revit there is a layer in the wall type definition called the "Core" layer. Each wall should have this layer, which can be at the center of the wall width or not. Refer to the following illustration:

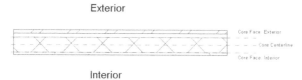

■ Other Location Lines are:

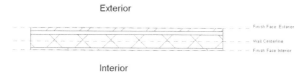

- If you want the drawing of the wall chained (the beginning of the new segment is the end of the last segment), then click on the **Chain** checkbox. If you click it off, you will specify two points for the first segment, two points for the second segment, and so on.
- Set the **Offset** value if the wall will be created with an offset value from the points selected.
- If the wall has an arc, click on the **Radius** checkbox and input the radius value.
- To control how two walls will join, select **Join Status** to be either Allow or Disallow:

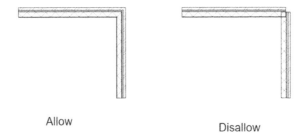

Allow                    Disallow

## WALL CONTROLS AND TEMPORARY DIMENSIONS

- When you click a wall, you will see controls appear over the wall, as shown in the following:

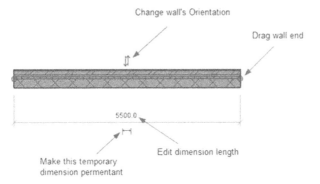

Change wall's Orientation

Drag wall end

5500.0

Edit dimension length

Make this temporary
dimension permentant

- If you click one of the walls in a chain of walls, you will see the following:

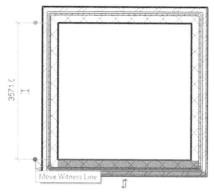

- The above example shows the measurement from inside to inside. If you click and hold the **Witness Line**, you can measure from center to center, or from outside edge to outside edge, or any combination of the above. Another way is to click the blue circle, which will move it from inside to the center to the outside of the wall.

- To input a permanent method, go to **Manage** tab, locate **Settings** panel, select **Additional Settings** button, and select **Temporary Dimensions** button:

- You will see the following dialog box:

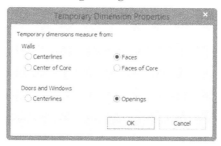

- Select to measure from the centerlines of the wall/core or the faces of the wall/core.

**NOTE** *To select chained walls, try the following:*

- *Move your mouse over one of the walls.*
- *Press [Tab] key, and all chained walls will be highlighted.*
- *Click to select them.*

- If you draw a series of walls and you want to close the shape, right-click, select **Snap Overrides**, and then select **Close**:

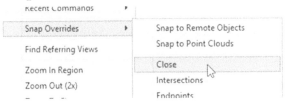

## EXERCISE 4-1   INPUTTING THE OUTSIDE WALL

**1.** Start Revit 2023.

**2.** Open the file **Exercise 4-1.rvt**

**3.** Make sure that you are at **00 Ground** view in Floor Plans.

**4.** Start Wall: Architecture command and input the following data:

   **a.** Wall Type = CW 102-85-100p (Exterior – Brick on CMU – Special)

   **b.** Base Constraint = Ground Floor

   **c.** Top Constraint = Parapet

   **d.** Location Line = Finish Face: Exterior

   **e.** Chain = yes

   **f.** Detail Level = Medium

**5.** Using Pick Lines tool, click the CAD file to import outside lines, making sure that the dashed lines are on the inside, ignoring the door openings, and not picking the curve.

**6.** Using Drag Wall End tool, close the door openings ignored in the previous step along with the entrance at the east elevation.

**7.** Using Wall family: Exterior Glazing and the Pick Lines tool, create the outside curve to be a Curtain Wall extending to the Roof minus 1250 (4'-2").

**8.** Go to the Roof view and close the gap between the curtain wall and the normal wall using CW 102-85-100p (Exterior – Brick on CMU – Special) if Revit gave you the following error: "The top of the wall is lower than the top of the wall – click Reset Constraints button."

**9.** Look at the model in 3D view; if the Brick is not on the outside, then click the Change wall's orientation control.

**10.** This is the final product of the exercise:

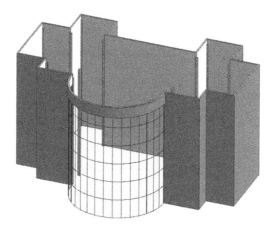

**11.** Save and close the file.

## ADDITIONAL MODIFYING COMMANDS

- In this part, we will continue looking at more Modifying commands to add interior walls.
- We will discuss the following commands:
  - Align command
  - Trim/Extend to Corner
  - Trim/Extend Single Element, and Multiple Elements
  - Split Element, and Split with Gap
  - Match Type Properties
  - Cut Profile

### Align Command

- This command will align two elements together.
- The first selected element will stay in its place, and the second will move to be aligned with the first one.
- To issue this command, go to **Modify** tab, locate **Modify** panel, and then click **Align** button:

- Do the following steps:
  - Pick the element that will stay in its place.
  - Then pick the element which will be aligned with the first one.
- Example:

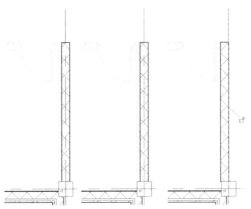

### Trim/Extend to Corner

- If you have ever used AutoCAD, this is similar to the Fillet command.
- This command will produce a neat intersection between two walls.
- Check the following two cases:

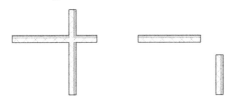

- You will receive the following result:

- To issue this command, go to **Modify** tab, locate **Modify** panel, and then click **Trim/Extend to Corner** button:

- The order of selecting is not important.

### Trim/Extend Single Element and Trim/Extend Multiple Elements

- These two commands are identical, except the first one will work with one element and the second command will work with multiple elements.
- Both commands can extend and/or trim depending on the case.
- First select the boundary element you will trim to or extend to.
- In case of trimming, select the part that will stay. In case of extending, select the end that you want to extend.
- To issue this command go to **Modify** tab, and using **Modify** panel, click one of the following two buttons:

### Split Element

- This command will split a single wall into two.
- There are two ways to use this command:
  - Click inside the wall to locate the position of the split.
  - Or, click **Delete Inner Segment** checkbox in Options bar. Then click twice on the wall, and Revit will delete the inner segment between the two clicks.

- To issue this command go to **Modify** tab, locate **Modify** panel, and then click **Split Element** button:

- In Options bar, you will see the following:

- The mouse will change to the following:

- Click either one click or two clicks as discussed previously.

## Split with Gap

- This command will split a wall into two parts, with a specified gap value between the two parts.
- This command is perfect for defining pre-cast concrete panels.
- The default value is 25.4 mm (which is 1").
- To issue this command go to **Modify** tab, locate **Modify** panel and then click **Split with Gap** button:

- Once you issue the command, Options bar will display:

- Click once on the desired wall and you will see the following:

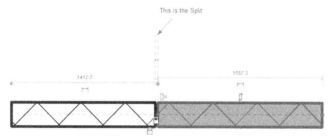

## Match Type Properties

- If you have two walls and you want one of them to match the properties of the other (type, height, etc.), then this is the right command for you.
- This command can work with walls, windows, doors, etc.
- To issue this command go to **Modify** tab, locate **Clipboard** panel, and then click **Match Type Properties** button:

- Do the following steps:
  - Start the command and the mouse pointer will change to the following:

  - Click the element that holds the desired type properties:

- The mouse pointer will change to the following:

- Click the other element that will change:

---

**NOTE** *To fillet a wall intersection with an arc, try the following, bearing in mind that we already have the wall intersection:*

- *Start the wall command.*
- *In **Draw** context tab, click **Fillet Arc** tool:*

- Click the first wall, then click the second wall, and finally, specify the arc radius, as shown in the following:

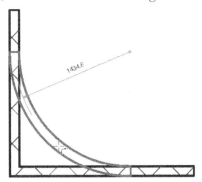

*NOTE* *By default, Revit will show all lines in the Floor / Ceiling plans in their real thickness, but you can choose to show them as thin lines. Refer to the following illustration:*

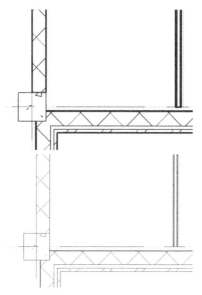

- To use this tool, go to the Quick Access Toolbar at the top left of the screen, and click **Thin Lines** button to switch it on/off:

## Create Similar

- If you have a wall carrying all the right information, and you want to create a new one similar to it, this command will help you do that easily.
- Click the existing wall, right-click, and then select **Create Similar** option:

- You can start inputting the wall right away, as it holds all the right information.
- You can use this command with doors, windows, and so on.

### Cut Profile

- When you add the outer wall with an inside finish layer of drywall (gypsum board), then you create an intersecting wall for an elevator or staircase, and you don't want this specific layer to be continuous in these two parts of the building.
- You need to stop the finishing layer in the staircase or elevator.
- To do that, you will use Cut Profile command.
- Do the following steps:
  - Disjoin the intersecting wall by selecting it, right-clicking the blue circle, and select **Disallow Join** command.

  - You will see the following:

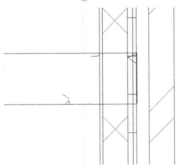

  - Start the Cut Profile command by going to **View** tab, locating **Graphics** panel, and clicking **Cut Profile** button:

- Select the layer you want to stop. It will turn orange. In one of the sides, draw a line representing the thickness of this layer. A small blue arrow will appear pointing to the direction you want to keep (if the initial one is not desired, click it to reverse the direction).
- Click (✔) to end the command. You will encounter the following:

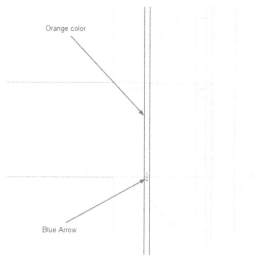

### Underlay

- If you go to the First floor plan, you will normally see the details in the lower floor, because the Underlay = Ground floor plan by default.
- You can switch this feature off using the properties of the floor plan you are in right now. Do the following:
  - Make sure you are not selecting anything (press [Esc] twice).
  - In Properties, you will see the properties of the current view. Under **Underlay**, set **Range Base Level** to None:

## EXERCISE 4-2    ADDITIONAL MODIFYING COMMANDS

**1.** Start Revit 2023.

**2.** Open the file **Exercise 4-2.rvt**.

**3.** Go to 00 Ground view in Floor Plans.

**4.** Show Thin Lines if it is turned off.

**5.** Zoom to the top right part of the building (ignoring all openings) and do the following:

    **a.** Start the Wall command and pick wall type: Elevator & Staircase – 200mm Concrete (Elevator & Staircase - 8" Concrete).

    **b.** Set the Base constraint = Ground Floor, Offset = 0.

    **c.** Set Top constraint = Roof, Offset = 0.

    **d.** Set Location Line = Wall Centerline.

    **e.** Using Line from the **Draw** context panel, draw the following walls starting from the midpoint of the column:

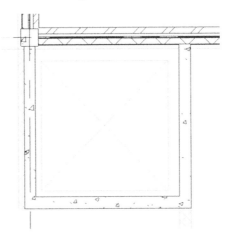

    **f.** Using Create Similar, draw the staircase walls (make sure to touch the outside wall).

**g.** You should receive the following result:

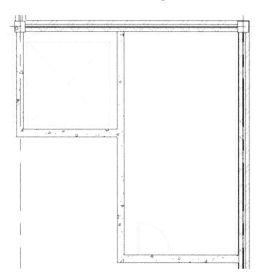

Using the Align command, align the left edge of the left wall of the elevator with the left edge of the column:

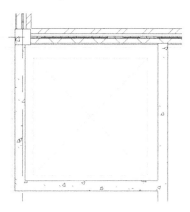

**6.** Using Cut Profile command, remove the finishing layer (drywall) from the staircase.

**7.** Do the inside walls (except the toilet inside walls) using the following, ignoring all the openings:

    **a.** Wall type = Interior – Blockwork 100 (Interior - 4" Masonry with Gypsum)

    **b.** Base = Ground Floor, Offset = 0

**c.** Top = First Floor, Offset = -300 (-25") (to accommodate for above floor)

**d.** Use the Pick Line tool

**e.** Location Line = Finish Face: Exterior

**f.** Follow the AutoCAD import

**8.** To do the partition inside the toilets, do the following:

**a.** Type = Interior – 66mm Partition (Interior - 2 1/8" Partition Special)

**b.** Base = Ground Floor, Offset = 0

**c.** Unconnected, Height = 2500 (8')

**d.** Use Pick Line and pick the lines of the CAD Import. Make sure that all walls are connected to the outside wall of the building.

**9.** Go to 01 First view floor plan, under Underlay set Range: Base Level = None. Using the CAD import, draw all interior walls (and toilet interior walls) using the same families and same settings we used in the 00 Ground view floor plan, except the lines representing the front wall of the offices (we will use a curtain wall in Chapter 6 to create them).

**10.** Save and close the file.

## WALL PROFILE

- Revit allows users to edit the wall profile.
- Editing the wall profile means you can change the wall look in an elevation.
- To edit the profile, do the following steps:
  - Go to the elevation view where your wall resides.
  - Pick the desired wall by clicking.
  - A context tab will appear; click **Edit Profile** button:

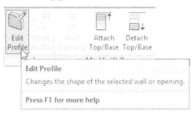

- The wall will change to having a ***magenta*** border.
- Make the necessary changes but make sure that your profile is always closed, continuous, and with no overlapping lines.
- Once done, click the (✓) in the context tab to end the profile editing.

**NOTE** *A closed area inside the profile will mean an opening.*

## HIDE AND UNHIDE ELEMENTS

- Sometimes Revit displays a significant amount of information in the Floor plan, Ceiling Plan, and Elevation Section views that may cause some confusion.
- You have the ability to hide some of the elements at any time during project development, then unhide them later.
- To hide any element, try the following:
  - Select one element of the desired element (say we need to hide the gridlines).
  - Right-click and you will see the following menu:

  - Select **Hide in View**, and then select either Elements or Category. If you select Elements, only the selected elements will be hidden. If you select Category, then all elements of this category will be hidden in this view.
- To Unhide any hidden element, do the following steps:
  - Click Reveal Hidden Elements button (the button showing a light bulb) in the View Control bar:

- You will see a red frame around the screen. Any hidden element will be shown in a red color. Select the desired elements to be unhidden, then right-click, and you will see the following menu:

- Click Reveal Hidden Elements button to end the process.

## EXERCISE 4-3   WALL PROFILE

**1.** Start Revit 2023.

**2.** Open the file **Exercise 4-3.rvt**.

**3.** Go to 00 Ground view floor plan.

**4.** Zoom to the right part of the building. The entrance at the east side of the building extends from the Ground Floor to the First Floor. We need to make an opening for this wall using the East Elevation view.

**5.** Go to East Elevation view.

**6.** From the top, click the wall at the middle. Refer to the following illustration:

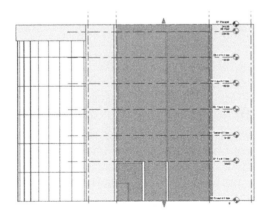

**7.** Click Edit Profile and edit the profile to look like the following from the bottom (no change at the top).

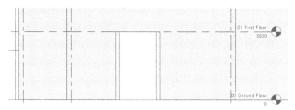

**8.** Looking at it by using the 3D view, you should receive the following result:

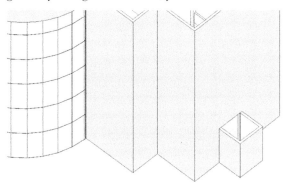

**9.** Go to 00 Ground floor plan and hide both the CAD Import and the gridlines (to select the CAD Import, hover over the text, and once you see the blue frame, click to select).

**10.** Zoom to the same place; you will discover that the outside wall is not displayed in the Ground Floor plan because we changed the profile of the wall. You should receive the following result:

**11.** Save and close the file.

## SLANTED WALLS

- This is the feature that everybody was anticipating.
- The only way to create a slanted wall before Revit 2021, was using Masses; which is a lengthy and complicated procedure.
- This is a very straight forward method. Do the following steps:
  - Draw the wall as you do always.
  - Select the desired wall.
  - Using Properties, under Constraints, located Cross-Section, change the value from Vertical to Slanted.
  - Set the Angle from Vertical value, normally positive value means to the inside, and negative value means to the outside.
  - Check the following image:

## SNAP BETWEEN 2 POINTS

- That can help in all commands but specially for Wall drawing
- If you want to start a new wall in middle distance between two columns, this feature will solve the problem, do the following
  - Start the wall command
  - To specify the first point, right-click, select Snap Overrides, select Snap Mid Between 2 Points option

- Specify the first point, specify the second point, then specify the first point of the wall
- Complete the wall command

## EXERCISE 4-4    SLANTED WALL

**1.** Start Revit 2023.

**2.** Open the file **Exercise 4-4.rvt**.

**3.** You are in Level 1 floorplan view

**4.** Select the wall at the right, and set the Cross-Section to Slanted, and set the angle to be 10°

**5.** Select the wall at the left, and set the Cross-Section to Slanted, and set the angle to be -10°

**6.** Look at your model in 3D view

**7.** Go back to Level 1 floor plan

**8.** Click the upper horizontal wall, right-click, select Create Similar, right click again, select Snap Overrides, select Snap Mid Between 2 Points option

**9.** Specify the first point to be the mid point of the left column, and the second point to be the mid point of the right column, go down a draw a wall 5000mm (16')

**10.** Press [Enter] to end the command

**11.** Save and close

## TAPERED WALLS

- You can create a wall with different top and bottom width by setting the selected wall to Tapered and manipulate the angle for the inner and outer side of the wall
- One of the wall layers must be ***Variable*** in order for Revit to allow you to set the Tapered angles
- If you want to create a tapered wall, do the following steps:
  - Draw the wall as vertical wall
  - Select it, from Properties click Edit Type button, and Duplicate it
  - Click Edit button beside Structure and set one of the layers as Variable

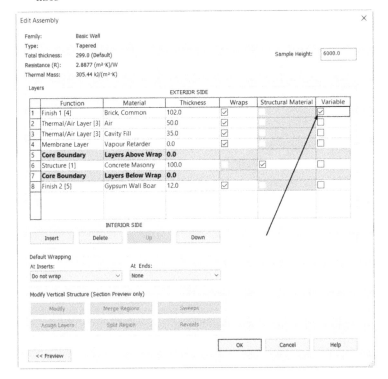

- Click OK to end the command

- Under Cross-Section Properties, set the Default Interior and Exterior angle, and Width measured at (Top, Bottom, or Both):

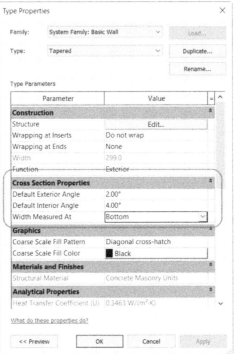

- Click OK to end the command
- While the wall is still selected, using Properties, change the Cross-Section to Tapered. Revit will use the settings in the Type Properties dialog box. You can turn on Override Type Properties, and set an instance Exterior and Interior Angles:

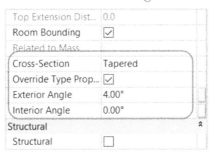

- Now, you can obtain vertical, slanted, and Tapered walls in Revit

## EXERCISE 4-5    TAPERED WALL

**1.** Start Revit 2023.

**2.** Open the file **Exercise 4-5.rvt**.

**3.** If the views are not tiled, tile them to see 3D, Level 1, and Section 1

**4.** Select the wall from any of the three views

**5.** Click Edit Type, and Duplicate it, and call the new type Tapered

**6.** Click Edit besides Structure

**7.** Set the layer at the top of the table (i.e. Brick, Common) to be Variable

**8.** Set the Default Exterior Angle = 4°, leaving Default Exterior Angle = 0, and Width Measured At = Top

**9.** While the wall is still selected, from Properties, set Cross-Section = Tapered, what is the result: _____
(it took the vlaues in the Type Properties)

**10.** From Properties, turn on Override Type Properties, and set the Exterior Angle=6°, What happened:_____
(the new values override the Type Properties)

**11.** Save and close

# NOTES

## CHAPTER REVIEW

**1.** Wall families are:

    **a.** System families

    **b.** Component families

    **c.** Component families that can be loaded to the project

    **d.** None of the above

**2.** Wall Centerlines and Core Centerlines are _____.

**3.** You can set the wall height to be Unconnected without specifying the height value:

    **a.** True

    **b.** False

**4.** Wall Profiles:

    **a.** Are a Magenta color

    **b.** Should be closed

    **c.** Should have no gaps or overlapping lines

    **d.** All of the above

**5.** You can hide and unhide elements in Revit Architecture:

    **a.** True

    **b.** False

**6.** Delete Inner Segment is an option that appears in the _____ command.

## CHAPTER REVIEW ANSWERS

**1.** a

**3.** b

**5.** a

## CHAPTER 5

# INSERTING DOORS AND WINDOWS

## This Chapter Contains

- Inserting doors and windows using pre-loaded families
- Loading door and window families
- Customizing door and window sizes

## INTRODUCTION

- In this chapter, we will learn how to insert doors and windows in Revit.
- Walls are the hosts for both doors and windows.
- We will discuss different techniques to insert both doors and windows:
  - Use the pre-loaded families.
  - Load the RFA file from the hard disk.
  - Customize the door and window sizes.
- We will discuss how to place doors and windows exactly using temporary dimension and Aligned dimension.
- Finally, we will discuss the door and window tags.

## INSERTING DOORS AND WINDOWS
## USING PRE-LOADED FAMILIES

- ▪ To insert a door or window, try the following:
  - • Go to the desired view. Though you can add doors and windows in many views, we prefer always to use Floor Plan views.
  - • Go to **Architecture** tab, locate **Build** panel, and click **Door** or **Window**:

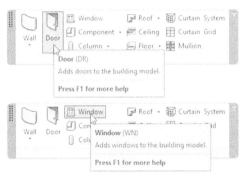

  - • In both commands you will see the following context tab:

  - • This context tab allows you to include a tag. Also, it will help you to load a non-pre-loaded family, and finally, it will allow you to create a new door/window using Model In-place.
  - • In Options bar, you will see the following:

  - • You can insert a tag in two orientations, either horizontal or vertical, with or without a leader (you have to specify the length of the leader). Refer to the following illustration:

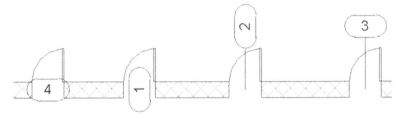

- Use Properties to choose the desired family and the desired size:

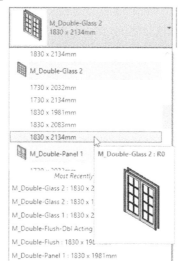

NOTE *In Revit templates, the default tag for doors is instance parameter (i.e., sequential); the next door will take the next number. But for windows, the default tag is type parameter (i.e., non-sequential), which means the same window type will always take the same number. In the coming chapters, we will learn how to load other types of tags for both doors and windows.*

## LOADING DOOR AND WINDOW FAMILIES

- Since door and window families are component families, you can load them to your current project.
- Door and window families are RFA files. Revit includes many families, which reside in the hard disk of your machine.
- Another way is to search the door and window families online free of charge, download them, and then copy them to the desired folder.

- To load the desired door and window family, do the following steps:
  - Go to **Insert** tab, locate **Load from Library** panel, and click **Load Autodesk Family** button:

- You will see the following:

- From the left pane, click Doors (or Windows) you will see the following:

- It will show all the doors in the Autodesk library. You can click the sub-category results click the names under Doors (or Windows).
- At the top right, click the shape of the globe to change language and the region of the content you would like to use:

- You can search by typing the phrase to filter your search
- Once you load the new door / window family, start the Door or Window command, and you will find the new family listed in the Type Selector

**NOTE** *You can load several RFA files in one shot. Use the check box at the top left of each family.*

## CUSTOMIZING DOOR AND WINDOW SIZES

- If you like the door family, but you need different size, you can create new sizes by creating duplicates of your type.
- To do that follow these steps:
  - Start **Door** or **Window** command and select the desired family.

- At Properties click the **Edit Type** button:

- You will see the following dialog box:

- Do not make the mistake of changing the values now; always create a duplicate and do your editing on it. Click **Duplicate** button:

- Type the new name of your type (normally, the name is the size of the door or the window).
- Click OK to end the creation process.
- Change the values you wish to edit.
- To finish the whole customization process, click OK.

## SPECIAL TECHNIQUES FOR DOORS AND WINDOWS

- To position a door or window precisely or after inserting, use the temporary dimensions. Refer to the following illustration:

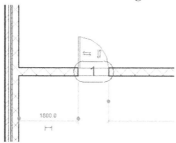

- Use **Aligned Dimension** tool to position multiple doors or windows exactly. Check the following example:
  - Notice the windows are not equally distributed over the distance between the staircase and the inside wall:

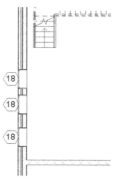

  - Start **Aligned Dimension** tool from Quick Access Toolbar at the top left of the screen:

  - At Options bar, change Wall centerline to Wall faces:

- Input the dimensions starting from the edge of the stair at the top, to the center of each window, ending it at the edge of the wall:

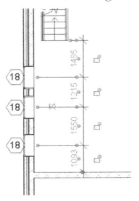

- There is a small icon of the letters **EQ** with a slash to indicate that the dimensions are currently unequal; click it once to make them equal:

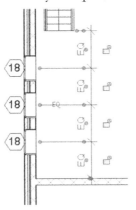

- To show the real distance, choose Properties and select the dimension (if not already selected); under **Other**, select **Value** for the **Equality Display** field:

**Reveal Constraint**

■ You can review your constraining process by highlighting the existing one.

■ Using the bar at the bottom, switch on the **Reveal Constraint** button.

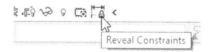

■ You can flip the door opening while inserting by pressing the Spacebar.

■ After inserting doors and windows, use the controls to flip them:

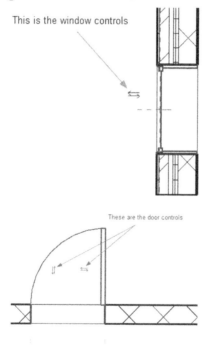

■ To copy the doors and windows from one level to other levels, do the following steps:
  • Select the desired doors and/or windows.
  • Use context tab titled **Modify|Doors** or **Modify|Windows.**
  • Locate **Clipboard** panel.

- Click **Copy to Clipboard** button:

- Click the pop-list of **Paste** button to see different options:

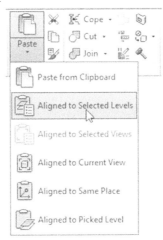

- Select **Aligned to Selected Levels** option. You will see the following dialog box:

- Select the desired levels you want to copy to, then click OK.

NOTE    *If you include door tags or window tags in the selection, the **Aligned to Selected Levels** option will not work.*

## EXERCISE 5-1    INSERTING DOORS AND WINDOWS

**1.** Start Revit 2023.

**2.** Open the file **Exercise 5-1.rvt**.

**3.** Go to **00 Ground** floor plan view.

**4.** Unhide the CAD Import and the gridlines.

**5.** Zoom to the top left office.

**6.** Start Door command, and there is only one family, which is M_Single-Flush; pick the size = 0915 × 2134mm (36" × 84"), and place it in the designated place as the CAD Import, making sure that there is a 200mm (2'-2") space between the door and the column.

**7.** Change the tag number to 01.

**8.** Do the same to the Cafeteria, Ladies' toilet, and Gentlemen's toilet doors by using the numbers 02, 03, and 04, respectively.

**9.** Do the same for the office at the right and the staircase.

**10.** Load   M_Door-Exterior-Double-Two_Lite.rfa   (Door-Exterior-Double-Two_Lite.rfa) door family from the Doors/Commercial folder, and insert size = 1800 × 2100mm (72" × 84") to the left of the curtain wall, as the CAD Import indicates.

**11.** Use the Aligned Dimension tool as follows to center the left door:

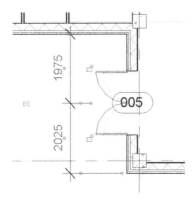

**12.** Click EQ to get equal distance. Show the value (it should be 2000, or 6'-6 ¾").

**13.** Delete the dimension (read the Warning message, click OK).

**14.** Insert the same door at the right side. Using the Align command, align the right door to be like the left door, then click the lock to close it.

**15.** Use Reveal Constraint button to look at the two constraints you just added, then turn it off.

**16.** Using Door command, select M_Single-Flush 0762 × 2032mm (30" × 80").

**17.** Click **Edit Type**, then click **Duplicate**, and name the new one 600 × 2000mm (26" × 76"). Change the Height to 2000 (76") and the Width to 600 (26").

**18.** Insert this new size on the Ladies' toilets using the temporary dimensions; insert them in the middle of each wall and you should have the following result:

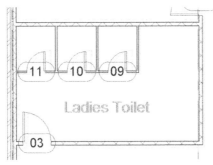

**19.** Do the same for the Gentlemen's toilet.

**20.** Load door family M_Door-Exterior-Revolving-Full Glass-Metal.rfa (Door-Exterior-Revolving-Full Glass-Metal.rfa) from the Doors/Commercial folder. Insert the size 2400 × 2400mm (96" × 96") in the right entrance as shown in the following:

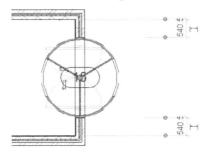

**21.** Copy all doors (except the three outside doors) to the 00 First floor plan (make sure you didn't select door tags).

**22.** Go to the 01 First floor plan.

**23.** Add  M_Door-Passage-Double-Flush.rfa  (Door-Passage-Double-Flush. rfa) 1800 × 2100mm (72" × 84") from the Doors/Commercial folder to the reception area.

**24.** Start the Window command; load the file M_Window-Sliding-double.rfa (Window-sliding-Double.rfa), then insert the size 1800 × 1200mm (60' × 48") at the left Manager's office. Change the tag to 01. It will look similar to the following:

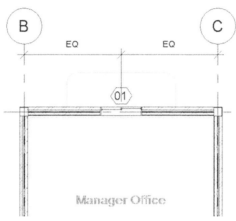

**25.** Do the same for the right Manager's office.

**26.** Start the Window command and load M_Window-Fixed-Arch-Top.rfa (Window-Fixed-Arch-Top.rfa); the size is 1800 × 2000mm (72" × 45").

**27.** Insert it as shown below and change the tag to 02:

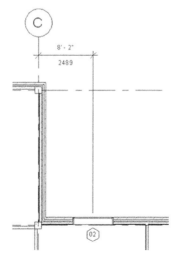

**28.** Insert three more, create the Aligned dimension in the following way, and click EQ. You should receive the following result:

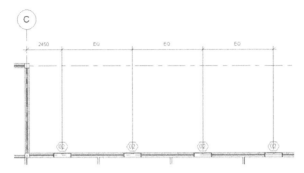

**29.** Copy all partitions, doors, and windows in the 01 First floor plan to the other four stories.

**30.** Go to the 3D view and look at your model; you should receive the following:

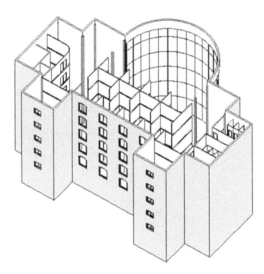

**31.** Save and close the file.

# NOTES

## CHAPTER REVIEW

1. To position doors and windows exactly at the desired place:

   **a.** Use the Aligned Dimension tool

   **b.** Use temporary dimensions before inserting

   **c.** Use temporary dimensions after inserting

   **d.** All of the above

2. Use _____ to flip a door opening while inserting.

3. If you select a door with its tag, you cannot copy it to other levels:

   **a.** True

   **b.** False

4. You can use doors and windows by:

   **a.** Using the pre-loaded families

   **b.** Downloading RFA files from online

   **c.** Customizing the sizes of an existing family

   **d.** All of the above

5. The Component family file is *.RVT.

   **a.** True

   **b.** False

6. You can choose in Aligned Dimension to measure from the center of the wall or from the face of the wall:

   **a.** True

   **b.** False

## CHAPTER REVIEW ANSWERS

1. d

3. a

5. b

## CHAPTER 6

# CREATING AND CONTROLLING CURTAIN WALLS

## This Chapter Contains

- All you need to know about curtain walls
- Controlling curtain grids
- Adding and manipulating mullions
- Adding storefronts
- Curtain panels
- Hiding and isolating elements
- Curtain Walls inside Slanted Walls

## INTRODUCTION

- There are three types of curtain walls in Revit Architecture:
  - Curtain Wall
  - Exterior Glazing
  - Storefront
- The Curtain Wall type is a single piece of glass and cannot be used in curved walls.
- In Chapter 4, we used Exterior Glazing in our model, which consists of glass panels only stacked over and beside each other.
- The Storefront type is aluminum sections and glass panels. This type has the power of being embedded inside Basic walls.

- In this chapter, we will discuss how to control the Exterior Glazing using Curtain Gridlines and Panels.
- Also, we will discuss how to insert and control the Storefront type.

## ALL YOU NEED TO KNOW ABOUT CURTAIN WALLS

- Any curtain wall is divided into:
  - Curtain Gridlines (which can be filled by a mullion or simply nothing)
  - Panels
- In order to select:
  - The whole wall, move your mouse to one of the outer edges as shown in the following (notice the tool tip), and then click to select:

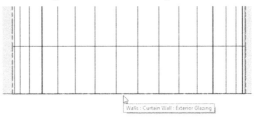

  - To select a panel, go to one of its edges and press [Tab] until you receive the following (note the tool tip) and then click to select:

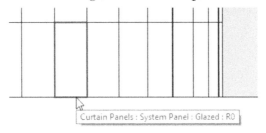

  - To locate a curtain gridline, simply move your mouse pointer to the desired line; it should highlight right away, but if it didn't, press [Tab] until you see the following:

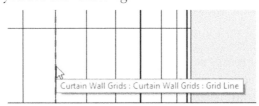

- The keyword here is curtain gridline:
  - Curtain gridlines can be replaced by mullions, but note that whenever you move the curtain gridline, the mullion will move with it.
  - When you move a curtain gridline, one of the panels will be smaller and the other will be bigger; hence, you can control the size of the panel through the movement of the curtain gridlines.
  - As an alternative method, you can add new curtain gridlines to the existing curtain wall, and new panel sizes will result. You can delete whatever you added but not the original curtain gridlines. To do that you need to **unpin** them first.
  - By default, Panels are holding Glass material, but you can assign any material you desire.

## CONTROLLING CURTAIN GRIDLINES

- Go to an elevation view that can show the curtain gridlines clearly.
- Hide anything may hinder your work.

### Moving Existing Curtain Gridlines

- To move a curtain gridline, you need to select it first.
- If it was pinned, then unpin it (to unpin anything pinned, use **Modify** context tab, locate **Modify** panel, and click **Unpin** button):

- There are several ways to move a curtain gridline:
  - Use **Move** command.
  - Hover your mouse over the curtain gridline until the Move icon shows, then click and move it.
  - Use the temporary dimension.
  - Use the Align command.

### Adding New Curtain Grid Lines

- This command will add new curtain gridlines.

- To issue this command, go to **Architecture** tab, locate **Build** panel, and click **Curtain Grid** button:

- The following context tab will appear:

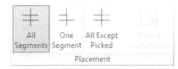

- If you hover over a horizontal curtain gridline, Revit will understand that you will add a vertical one.
- If you hover over a vertical curtain gridline, Revit will understand that you will add a horizontal one.
- When you want to add a new curtain gridline, select from the following choices:
  - Whether the new curtain gridline should cut all segments
  - Whether the new curtain gridline should cut only the picked segment
  - Whether the new curtain gridline should cut all segments, then pick some undesired segments out

### Deleting Curtain Grid Lines

- To delete a curtain gridline:
- Simply select it and press [Del] at the keyboard.
- You cannot delete an original curtain gridline—only user-added ones.

### Controlling Curtain Grid Lines Using Properties

- When you select the whole curtain wall, two parts of the Properties will allow you to control vertical and horizontal grids as shown in the following:

- Control the Justification of the vertical and horizontal grid lines:
  - **Beginning** – means starts from the bottom (if it was horizontal), or left (if it was vertical)
  - **End** – means starts from the top (if it was horizontal), or right (if it was vertical)
  - **Center** – means starts from the center of the distance
- You can select to change the Justification with or without an Offset value

### Add/Remove Segments

- This option will be available only when you select a curtain grid line. You will see the following context tab:

- This command allows you to pick any segment of a selected curtain grid line and delete it if it exists or add it if it is not there.

---

**NOTE**

---

- *If you have a curved curtain wall, the best place to add/remove segments in the 3D view.*
- *This could help you remove an original curtain grid line by removing all of its segments (but the curtain grid line will stay).*

## ADDING AND CONTROLLING MULLIONS

- You can replace curtain grid lines with mullions.
- By default, mullions are:
  - Different shapes and sizes (Rectangular, Square, L-Shape, Trapezoidal, Circular, etc.)
  - Material used is always Aluminum (but you can change it)
- To issue this command, go to **Architecture** tab, locate **Build** panel, and click **Mullion** button:

- You will see the following context tab:

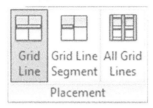

- You can select one of the three options:
  - **Grid Line** – you will add mullions covering the whole grid line
  - **Grid Line Segment** – you will add mullions for the selected segment
  - **All Grid Lines** – you will add mullions covering all grid lines in the curtain wall

### Selecting a Group of Mullions

- When you select a group of adjacent mullions, you will see the following context tab:

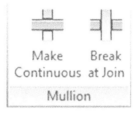

  - **Make Continuous** will make a group of selected mullions continuous. Refer to the following illustration:

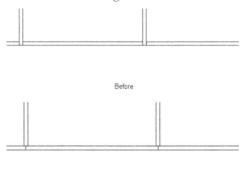

  - **Break at Join** will do the opposite.

## ADDING STOREFRONTS

- A storefront is special curtain wall which consists of mullions and glass panels.
- This special curtain wall has the ability to penetrate basic walls and occupy their space.
- This is a wall type, hence everything we know about walls apply here. You need to add in the middle of the other walls, as in the following:

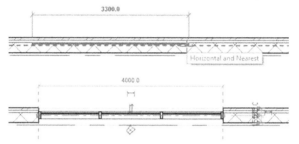

- This how it looks in elevation view:

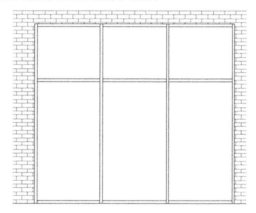

## CURTAIN PANELS

- By default, all panels in any curtain wall are presented as glass panels.
- You can do two of the following:
  - You can replace a panel with special doors or windows.
  - You can change the material of a panel.

### Replacing Panels

- To replace a panel with doors and windows, try the following:
- Load the special door or window family, but do not use it in the normal way of inserting doors or windows. End the door or window command.
- In elevation view, select the desired panel then Unpin it.
- Go to Properties and click the arrow beside the current type (System Panel Glazed). Check the category called **Most Recently Used Types**, find the loaded door or window among them, and select it:

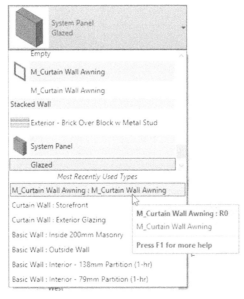

### Panel Material

- To change panel material, try the following:
  - In elevation view, select the desired panel then Unpin it.
  - Go to Properties and click **Edit Type** button. Click **Duplicate** button and give a name for the new System Panel.
  - Locate **Material** field:

- When you click **Glass** field, at the far right click the small button with three dots to show **Material Browser** dialog box:

- Browse for your new material or use the top edit box to type the name of the new material to search for it in the library. Either way, select it and click OK, then click OK to end the creation of the new System Panel.

**NOTE**

- *When you select gridlines, mullions, or panels, you will not see the controls nor the temporary dimensions.*
- *If you want to see them, using Quick Access Toolbar, locate and click Activate Controls and Dimensions.*

## HIDING AND ISOLATING ELEMENTS

- This is another method to hide elements or categories of elements, along with the function of isolating elements or categories of elements (isolating means the selected elements will be shown, others will be hidden).

- Select the desired element.
- At the View Control Bar select **Temporary Hide/Isolate** button:

- You will see the following menu:

- Pick between Hiding or isolating elements or categories.
- You will see a cyan frame around your screen with title Temporary Hide/Isolate to remind you of what you did.
- To reset back to normal, click **Temporary Hide/isolate** button, and select **Reset Temporary Hide/Isolate**.

## CURTAIN WALLS INSIDE SLANTED WALLS

- In Chapter 4, we introduced the Slanted Walls.
- In this chapter, we introduced the Storefront, which can be embedded within a vertical wall.
- Curtain walls cannot be embedded in slanted walls.
- The best way to handle this case is to cut the basic wall, then add the curtain wall, finally set the curtain wall Cross-Section to Slanted and set the same angle you set for the basic wall.

## EXERCISE 6-1   CREATING AND MANIPULATING CURTAIN WALLS

**1.** Start Revit 2023.

**2.** Open the file **Exercise 6-1.rvt**.

**3.** *Hint*: In order to see the pins of the curtain wall click Activate Controls and Dimensions from the Quick Access Toolbar.

**4.** Go to South elevation view. Hide all grid lines.

**5.** Unpin all horizontal curtain grid lines.

**6.** Using Align command, align all horizontal lines to level lines.

**7.** Add three new curtain grid lines beneath the line at the First-Floor level, separated by 1200 (4'-6") each.

**8.** Assign mullions for all curtain grid lines using Circular Mullion 25mm (2.5").

**9.** Select one of the mullions and isolate the category.

**10.** Select all bottom mullions and make them continuous.

**11.** Do the same for the top mullions. Reset the Temporary Hide/Isolate.

**12.** Replace the center panel and the ones above it (look at the following illustration) above First Floor level with the window family M_Curtain Wall Awning.rfa (Curtain Wall Awning.rfa).

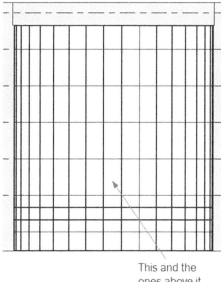

This and the
ones above it

**13.** Go to the 3D view and check how the panels holding windows look different from the other panels.

**14.** Go to the Ground Floor view.

**15.** Add a Storefront in the north side of the building between gridlines C & D (as shown in the following). Storefront specification is as follows:

   **a.** Base Constraint = Ground Floor

   **b.** Base offset = 0

   **c.** Top Constraint = First Floor

   **d.** Top offset = -2000 (-6')

   **e.** Length = 10000 (33')

   **f.** Distance from left grid line = 5000 (16'-6")

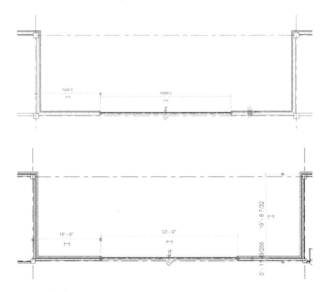

**16.** Go to North elevation view.

**17.** Select the horizontal curtain grid line, unpin it, and then move it up to be 1000 (3'-0") from the top.

**18.** Load the door family M_Door-Curtain-Wall-Double-Storefront.rfa (Door-Curtain-Wall-Double-Storefront.rfa).

**19.** Replace the panels (after unpinning them) with the three doors as shown in the following with the new loaded door family:

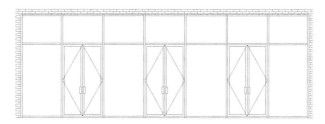

**20.** Delete the mullions beneath each door.

**21.** Change the top panels to show Aluminum material (create a duplicate and name it ***Alum Cover***).

**22.** Go to the 3D view, look at the north elevation, and check the different material of the entrance.

**23.** Go to the Ground Floor view.

**24.** Zoom to the right wall of the curved curtain wall, and add a 2000 (6'-0") measured from left grid line 2750 (9'-0") Storefront extending from the Ground Floor up to the Roof -2000 (-6'-0"), as in the following illustration:

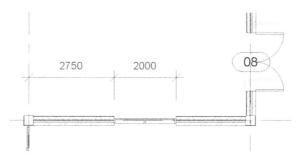

**25.** Do the same for the left part of the building.

**26.** Make the storefront mullions on both sides aligned with the curved curtain walls exactly (start from bottom to top – Unpin all grid lines in the right and left storefront walls).

**27.** Change the profile of the two storefronts to look similar to the following illustration:

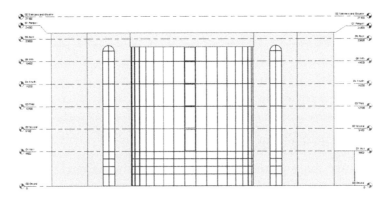

**28.** Go to the 01 First floor plan view.

**29.** Show the CAD import.

**30.** Zoom to the north part of the building which contains four offices.

**31.** Add a Storefront covering the front of the office at the right (make sure the curtain wall touches the right and left walls) Base = 01 First, Top = 02 Second -300mm (-25").

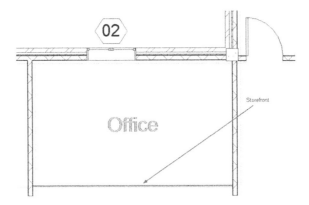

**32.** Go to Elevation 1-a.

**33.** Do not change the locations of the mullions, just add the M_Door-Cur-tain-Wall-Single-Glass.rfa (Door-Curtain-Wall-Single-Glass.rfa) door as designated below:

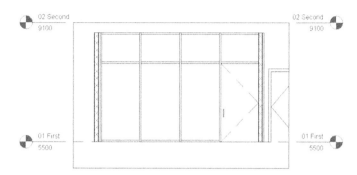

**34.** Using the copy and mirror commands, complete the rest of the offices (change the opening of the doors when needed).

**35.** Save and close the file.

## NOTES

# CHAPTER REVIEW

1. The keyword in working with curtain walls is the curtain grid line:

    **a.** True

    **b.** False

2. One of the following statements is NOT TRUE:

    **a.** You can replace curtain wall panels with special doors and windows made specifically for curtain walls.

    **b.** You can change the material of any curtain wall panel.

    **c.** You can replace curtain wall panels with any door and window family.

    **d.** Mullions come with multiple shapes and sizes.

3. You cannot delete any original curtain grid line:

    **a.** True

    **b.** False

    **c.** True for inside curtain grid lines, but for the outside you can delete them.

    **d.** True for the outside curtain grid lines, but for the inside you can delete them.

4. Storefront wall types can penetrate any type of other walls:

    **a.** True

    **b.** False

5. _____ button will make a group of selected mullions continuous.

6. You may need _____ at the keyboard to select a panel in the curtain wall.

7. Before you control curtain grid lines _____ them first.

## CHAPTER REVIEW ANSWERS

**1.** a

**3.** a

**5.** Make Continuous

**7.** Unpin

## CHAPTER 7

# CREATING FLOORS

## This Chapter Contains

- Creating floors using two methods
- Joining elements
- Creating a shaft opening
- Creating a floor with slope

## INTRODUCTION

- In this chapter, we will learn how to create floors.
- Floors will be the tool to define balconies and decks.
- Create floors by selecting the bounding walls, or by sketching using drawing tools. Later on you can edit the profile just like we did with walls.
- We can create shafts to create an opening in a floor (or floors) depending on the height of the shaft.
- We will learn some extra functions related to walls.

## CREATING FLOORS BY SELECTING WALLS OR SKETCHING

- This command will add a floor using the bounding walls or by sketching using drawing tools.
- To start the Floor command, go to **Architecture** tab, Locate **Build** panel, and click the popup-list of **Floor** to select **Floor: Architecture**:

- In Properties palette, select the desired floor family. Floor families are like wall families; they are system families:

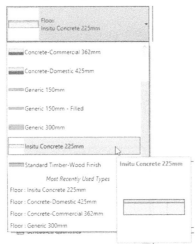

- The following context tab will appear:

- • The default option is to specify the Boundary Line.
  - • In the Profile Sketching mode, you either select the bounding walls (the default option), or start sketching using the drawing tools.
  - • Use **Slope Arrow** button to specify the slope of the whole floor.
  - • Use **Span Direction** button to specify the direction of the floor span. By default Revit will specify the direction of the floor span as the first picked line of the profile.
- ▪ Options bar will show:

- ▪ Specify whether the floor should be extended into the wall (to core), or not. You can also specify an offset value inside or outside.
- ▪ The profile should be closed with no overlaps.
- ▪ When done, click (✓) to end the command, and you may see the following message:

- ▪ Revit asks if you want to attach the lower walls to the upper walls or not.
- ▪ If you created part of the floor outside the walls (e.g., a balcony), Revit will show you the following message:

- ▪ The above message means Revit is asking to join the floor and the bounded walls and remove the overlapping volume from the walls.

- *Just like the Wall command, to create an opening in the floor, simply draw a closed shape inside the boundary lines.*
- *Floors are normally created at the level you are in. If you go to an elevation view, you will see the following. So the thickness of the floor will go below the level (this is of course by default):*

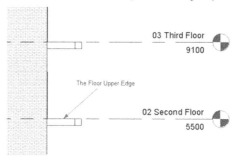

## JOINING GEOMETRY

- For typical floors, when you create a floor with balconies, the floor normally joins the wall, making an opening in the wall.
- But when you copy this floor for the other levels, the copied floors will not join the wall, so you should join them manually with the **Join Geometry** command.
- To join two elements together (here we are talking about floors and walls) do the following:
  - Go to **Modify** tab, locate **Geometry** panel, click the pop-up list of **Join** button, and then select **Join Geometry** button:

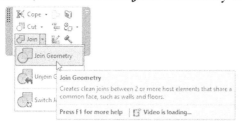

  - At Options bar, click **Multiple Join** if you want to join multiple elements and not only two:

  - Select the first element and then the second element.

## HOW TO CREATE A SHAFT OPENING IN FLOORS

- As mentioned earlier, you can create openings while sketching the floor profile.
- Also, after creating a floor you can create a shaft opening, specifying a Base Constraint and Top Constraint.
- The user should be in the floor plan to create a shaft.
- To do that, go to **Architecture** tab, locate **Opening** panel, and select the **Shaft** button:

- You will be inside a sketch mode to draw the plan of the shaft. The following context tab will appear:

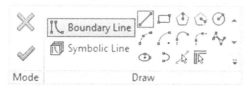

- In Options bar, you will see the following:

- Keep **Chain** checkbox on, so you can draw multiple lines. Specify an offset if the shaft has an offset to existing walls or elements. If the shape contains arcs, specify the radius of the arcs.

- Use **Symbolic Line** to draw lines to indicate a void in the plan and all the plans the shaft reaches.
- Here is a shaft Opening with Symbolic Lines:

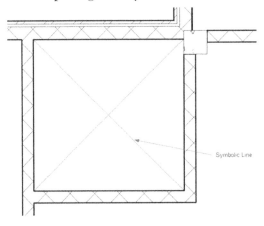

- Control height of the shaft using Properties:

- In the above example, the shaft extends from the Ground floor (-150mm) to the Roof level, penetrating all floors in between these two levels.

## CREATING A FLOOR WITH SLOPES

- You can create two types of slopes in floors:
  - Create a one-direction slope
  - Create a slope for drainage

**One-Direction Slope**

- You can specify a slope in one direction for a floor (or roof).
- To do that, follow these steps:
  - Select the desired floor.
  - At the context tab, click **Edit Boundary** button.
  - At the context tab, click **Slope Arrow** button:

  - Draw the slope arrow (the length and direction are very important), and you will receive the following:

  - The first point will be the tail and the last point will be the head.
  - When done, select the arrow and look at the Properties:

  - Under Constraints, click Specify, and you will have two choices, either Height at Tail or Slope.
  - If Height at Tail is selected, specify the level of the head and tail (the current is default, which means the current level). Also, specify the Height Offset at tail and head.

**Slope for Drainage**

- You can specify the slope for drainage in the toilets and kitchens using this feature.
- You will first need to add Split Lines, which will split the area of the toilet from the other parts of the floor.
- Then add a point with a lower elevation to create the slope.
- Depending on the floor family used, Revit will slope the entire floor or only the top layer.
- To create a drainage slope, do the following:
  - Select the desired floor.
  - You will see the following context tab:

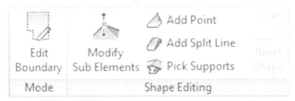

- Start with **Add Split Line** button to separate the area in which you want to create the drainage slope from the rest of the floor slab.
- Select **Add Point** button, then at Options bar, specify the desired value (it should be negative—assuming the other points to be at zero slope).
- You will receive the following:

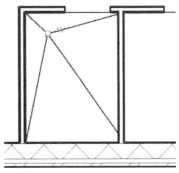

- Use the **Modify Sub Elements** command, to edit the split lines, point elevation and location, and folding lines. Once you are at the Modify Sub Elements, you will be able to convert any folding line to split line, using **Convert Lines** command.
- If it did not work out and you want to start over, do the following:
  - Select the desired floor.

- At the context tab click **Reset Shape** button and everything will return to the point before you specified any drainage slopes:

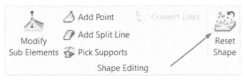

## UNDERLAY AND LINEWORK

- In Chapter 4, we discussed the Underlay feature, which will allow you to see the floor below you as an underlay.
- Linework will allow you to draw lines over the edges of other elements to highlight them.
- To start the command, go to **Modify** tab, locate **View** panel, and click **Linework** command:

- A context tab will appear; select the desired **Line Style**:

## EXERCISE 7-1    CREATING FLOORS

1. Start Revit 2023.

2. Open the file **Exercise 7-1.rvt**.

3. Go to 00 Ground floor plan.

4. Hide the CAD import and gridlines.

5. Start Floor command and select family Insitu Concrete 200mm (Insitu Concrete 8").

6. Make sure that Extend into wall (to core) checkbox is turned off, and Offset=0.

7. Pick the outside edge of all the outside walls one by one.

8. Press [Esc] twice to get out of the selecting mode.

9. Zoom to the right entrance.

10. If you did not pick the three walls, pick them right now, and fix the boundary to be continuous with no overlaps. When done click (✓) to end the command.

11. Go to the South elevation view to make sure that the floor is covering the whole ground floor up to the outside edges of the walls.

12. Go to the 01 First floor plan. Using the CAD import you will notice there are some extra lines to specify outside edges and inside edges for the floor slab.

13. Using the same floor family, do the boundary of the first floor making sure **Extend into wall (to core)** checkbox is turned on. (**Hint:** do the walls first, then do the other lines, and make them connected and closed.)

14. If the following message is produced: "Would you like walls going up to this floor's level to attach to its bottom?" answer No.

15. If the following message appears: "The floor/roof overlaps the highlighted wall(s). Would like to join geometry and cut the overlapping volume out of the wall(s)?" then answer Yes.

16. The new floor is already selected; copy it to the 2nd, 3rd, 4th, and 5th levels.

17. Go to the 3D view and make Visual Style = Hidden Line.

**18.** Notice that first floor slab penetrates the wall and you can see the edge lines. But from 2nd to 5th, the wall edges are not there! We need to join the two walls with the floor slabs. (***Hint:*** make sure it is set to Multiple Join before you start.)

**19.** Make a shaft opening for the elevator extending from the Ground Floor minus 500mm (-10") to the Roof and draw symbolic lines. Visit the 3D view to make sure it penetrates both sides.

**20.** Go to the Ground Floor.

**21.** Zoom to the office at the top left. Start the Floor command and select the Offices floor type. Using Properties, set Height Offset to Level = 20 (11/16"). Select the walls of the office from inside (turn off Extend into wall [to core]). Do the right office as well (if any message comes up, answer No).

**22.** Start the Floor command, select the Toilets & Cafeteria floor type, set the Height Offset level = 70 (3"), and do the Cafeteria and the two toilets (each one in a separate command).

**23.** In Properties, locate Underlay and make the following settings:

    **a.** Range: Base Level = 00 Ground

    **b.** Range: Top Level = 01 First

    **c.** Underlay Orientation = Look up

**24.** Using the Linework command, change the Line Style = <Overhead>. Select the edges of the above floor which is near the curtain wall and the two balconies.

**25.** Set Range: Base Level = None

**26.** Save and close the file.

## NOTES

## CHAPTER REVIEW

1. A Floor family is like a Wall family; both are System families:

   **a.** True

   **b.** False

2. The _____ button will open a hole in floors.

3. While you are drawing a Boundary line using Pick Walls, you can:

   **a.** Pick the outer edge of the wall

   **b.** Pick the inner edge of the wall

   **c.** Pick the Core edge

   **d.** All of the above

4. Use _____ to draw lines to indicate a void in the plan and all the plans the shaft reaches.

5. By default, Revit Architecture places the top edge of the floor aligned with the level:

   **a.** True

   **b.** False

6. _____ will allow you to draw lines over the edges of other elements to highlight them.

## CHAPTER REVIEW ANSWERS

**1.** a

**3.** d

**5.** a

## CHAPTER 8

# CREATING ROOFS

## This Chapter Contains

- Creating a roof by footprint
- Creating a roof by extrusion

## INTRODUCTION

- Roofs are very similar to Floors. Roofs will cover the last level in a building and cover any entrance.
- The Roof families are system families like floor and wall families.
- There are two types of Roofs:
  - **With Roof by Footprint**, you can create a flat roof or slope (Shed, Gable, and Hip) roof.
  - **With Roof by Extrusion**, you can create irregular shapes and extrude them to form a roof.
- Meanwhile we will discuss the Reference Plane tool, which will help us place the roof by extrusion in the right elevation.

## ROOF BY FOOTPRINT

- This command will allow the user to create a flat, hip, shed, or gable roof:

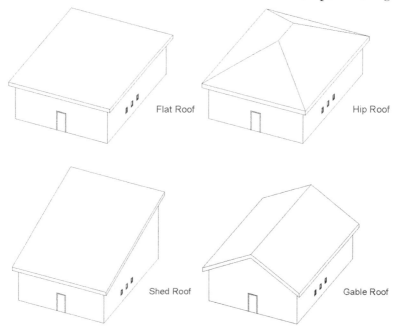

Flat Roof     Hip Roof

Shed Roof     Gable Roof

- You can specify the value of the slope (the default value = 30°).
- You can add an overhang distance.
- To start this command, go to **Architecture** tab, locate **Build** panel, click the popup-list of **Roof**, and then select **Roof by Footprint** button:

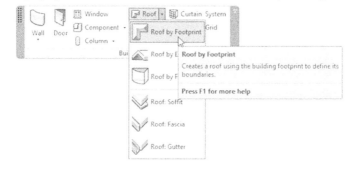

- If you are not in the right level, you will see the following message to help you move to the correct level:

- Select the desired roof family, you will see the following:

- The following context tab will appear:

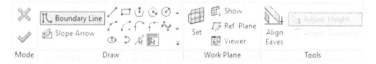

- The default option is **Boundary Line** with draw tools available. This option will allow you to select the desired roof bounding walls. We will discuss the other options in the upcoming pages.
- You will see the following in Options bar:

- Specify the following before you pick the walls:
  - Whether to define the slope for each selected wall (default value is 30°)
  - The value of the overhang distance
  - Whether or not to measure the overhang from the wall core rather than from the outer edge of the wall

- In the right view, and based on the settings you made, start clicking the desired walls. You will receive the following:

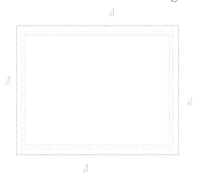

- In the above example, all four walls have an overhang and slope.
- While you are inside the roof command, press [Esc] twice to get out of the picking walls mode, then click one of the magenta lines, and you will see the following:

- You can edit the overhang distance and the value of the slope. You will see the two arrows to flip the boundary profile to the inside or to the outside.
- Click anywhere to remove the selection.

**Slope Arrow**

- You can mix and match between the slopes and the slope arrow.
- The slope arrow here is identical to the slope arrow applied to the floor, which we discussed previously.
- Select Slope Arrow button on the context tab:

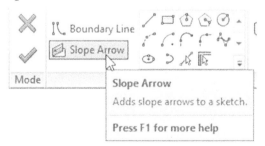

- The draw tool will appear similar to the following:

- You can draw a line or pick an existing line. You will receive the following:

- The length of the arrow and the angle are important factors.
- Select the arrow, look at the Properties, and you will see the following:

- Under Constraints click **Specify**, and you will have two choices, either Height at Tail or Slope.
- If Height at Tail is selected, specify the level of the head and tail (the current is default, which means the current level). Also, specify the Height Offset at tail and head.

- When done with the roof boundary profile, click (✓) to end the command.

### Attach the Top or Base of the Wall to the Roof

- Once you are done with a hip, shed, or gable roof, you need to attach the walls to the roof. Normally, you will have the following case:

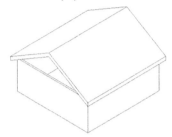

- To attach the walls to the roof, do the following:
  - Select the desired wall(s).
  - At the context tab, you will see the following:

  - Click the **Attach Top/Base** button.
  - Select the roof. This is the final product:

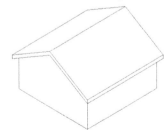

  - The same applies if you want to detach walls.

NOTE *To create a floor plan view and/or reflected ceiling plan view for levels that had neither view during creation, do the following steps:*

- Go to **View** tab, locate **Create** panel, click the popup-list of **Plan Views**, and then select **Floor Plan** button:

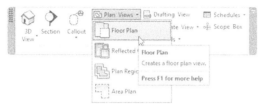

- You will see the following dialog box:

- Select the desired levels to create a floor plan for them and then click OK.

## EXERCISE 8-1   CREATING A ROOF BY FOOTPRINT

**1.** Start Revit 2023.

**2.** Open the file **Exercise 8-1.rvt**.

**3.** Go to the 06 Roof floor plan view.

**4.** Under Underlay, set Base: Range Level = None, and hide Gridlines.

5. Start the Roof by Footprint command and select family = Warm Roof – Concrete, Define Slope = Off, Overhang = 0, and Extend to wall core = On.

6. Use the Pick Wall tool to select all the external walls.

7. When the message appears, answer yes.

8. Go to the 3D view and look at your model.

9. Go to the 00 Ground floor plan and change the settings of the elevator shaft to Top offset = 150 (0'-6"). Go to the 3D view to make sure the shaft penetrates the roof.

10. Create a Floor Plan for the Parapet level.

11. Go to the Parapet level and make Underlay Range: Base Level = Roof.

12. Using the same Roof family, create a hip roof as shown in the following illustration (make each one in a separate command). Note the lower line of the boundary, which coincides with the outer edge of the parapet. The Overhang = 500mm (1'-8").

13. Go to the 3D view and you should have the following image:

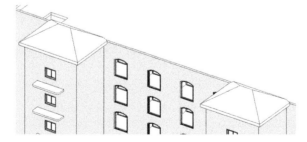

14. Go to the 01 First floor plan and hide the CAD import.

**15.** Under Underlay, set Base: Range Level = 00 Ground.

**16.** Zoom to the east entrance, and create a gable roof using roof family = Warm Roof – Timber, using the following illustration, Overhang = 500mm (1'-8"):

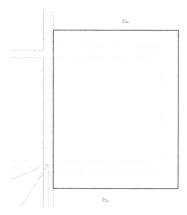

**17.** Select the three walls and attach them to the newly created roof.

**18.** You should have the following result:

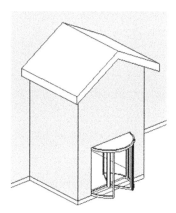

**19.** Save and close the file.

## ROOF BY EXTRUSION

- Roof by extrusion will allow you to create irregular shapes for roofs or entrance shades.
- Draw an **open profile** in one of the elevation sides, then Revit will extrude it perpendicular to the view you are in.
- You can control the real length of the extrusion.
- You can use a tool called **Vertical** to create something similar to the shaft.
- But since we will work in elevation view, and elevation views are tricky, we will use Reference Planes.

### Reference Plane

- Reference Planes will help you specify your working plane.
- Try the following:
  - Go to one of the floor plan views.
  - Go to **Architecture** tab, locate **Work Plane** panel, and click **Ref Plane** button:

  - The context tab appears as:

  - Options bar will appear as the following:

  - You can draw lines or pick lines with or without offset.

- Click "Click to name" to give your Reference Plane a proper name or use the Properties palette:

- The final product will resemble the following:

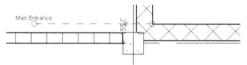

- To create a roof by extrusion, do the following:
  - Go to the elevation view in which you want to create the extrusion.

- Go to **Architecture** tab, locate **Build** panel, click the popup-list of **Roof** button, and select **Roof by Extrusion** button:

- Since we are working in an elevation view, Revit will show **Work Plane** dialog box:

- Use one of the three available methods. The first is to specify by **Name** (which is the method in the previous picture); if you already created and named a reference plane, you will see it at the end of the list. The second method is to **Pick a plane**, by selecting a wall for instance. Finally, the user can **Pick a line and use the work plane it was sketched in**.
- When done, click OK, and you will see the following dialog box:

- Specify the level that your roof should not exceed (it will appear as green dashed line) with or without an offset value (this is just a guide-line, you can go over it if you want).
- The context tab will resemble the following:

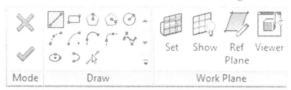

- Use any of the drawing tools to draw the open shape to be extruded.
- Use **Ref Plane** to create additional planes to draw your profile exactly.
- You will see the following:

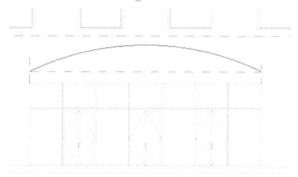

- If you are satisfied with the profile, click (✓) to end the command.
- This will be the final product:

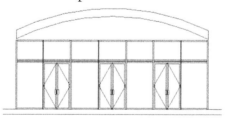

- Go to the **Site** view (this view is very special, where you can see all the details of the building regardless of the levels in which they were created).

- You will see the roof; click it to show the two handles at the beginning and end and move both to get the desired result. It will resemble the following:

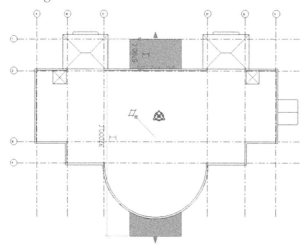

- While you are selecting the roof, the following context tab will appear:

- The **Vertical** option will create an opening in the roof. This is a very good tool to clean the parts inside the building when you create an outside shade.

NOTE *There is a special Window type called Skylight, which can be inserted inside any roof, even if it is inclined or even wavy.*

- You can load it from Load Families in the Window command.

## EXERCISE 8-2    CREATING A ROOF BY EXTRUSION

1. Start Revit 2023.

2. Open the file **Exercise 8-2.rvt**.

3. Go to the 00 Ground floor plan.

4. Zoom to the north entrance.

**5.** Create a Reference Plane as shown in the following and call it **Main Entrance**. Place it 3000mm (10'-0") away from the edge of the wall:

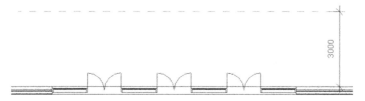

**6.** Go to the North elevation.

**7.** Zoom to the main entrance.

**8.** Start the Roof by Extrusion command.

**9.** When the Work Plane dialog box comes up, click the Name option, and then select Main Entrance from the list. Specify 01 First as your roof reference level with offset = 0.

**10.** Draw the following profile bearing in mind the following facts:

   **a.** Within the Roof by Extrusion command, draw three reference planes: two vertical flush with the left and right edges of the Storefront, and the horizontal 500mm (0'-20") above the top edge of the Storefront.

   **b.** Use the Start-End-Radius to draw the arc.

   **c.** Draw the arc from left to right picking the left point and right point. To specify the radius, move the mouse up until it touches the level of the first floor, then click. Press [Esc] twice to get out of drawing mode, select the arc, and specify the angle to be 80°.

   **d.** You should receive the following:

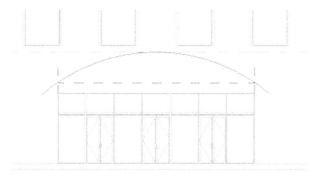

**11.** From the Properties, select the family to be Outer Shade, then click (✓) to end the command.

**12.** Go to the Site view in the floor plan.

**13.** From the south side, you can see the edge of the roof. Select it and move it to appear as the following:

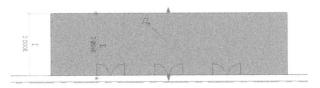

**14.** Go to the 3D view and look at your model; you should receive the following result:

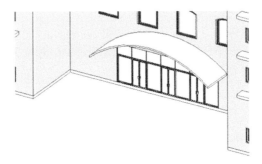

**15.** Go to the Roof floor plan, and under Underlay, set Range: Base Level = None.

**16.** Create a reference plane like the following and call it Staircase End. Place it 300 mm (1'-00") from the outer face of the wall hosting the door:

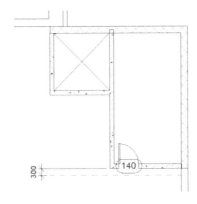

**17.** Go to South elevation and zoom to the walls of the staircase and elevator on the roof.

**18.** Start the Roof by Extrusion command.

**19.** Select Name, then select Staircase End, and then select 08 Staircase and Elevator level with offset = 3000 (10'-0").

**20.** Using the Reference Plane command inside Roof by Extrusion, create the following reference planes above the staircase and elevator:

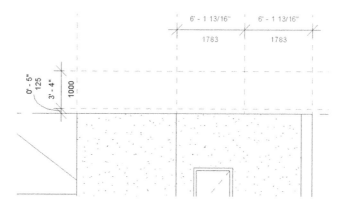

**21.** Now draw the profile like the following:

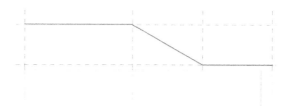

**22.** Select roof family = Generic – 125mm (Generic – 5").

**23.** Click (✓) to end the command. This is the final product:

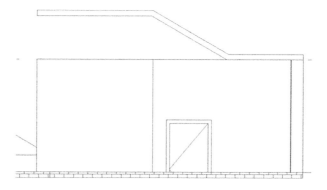

**24.** Go to the Site view, and correct the roof to be aligned from the top to the outer edge of the main wall, and from the bottom to be aligned with the reference plane, as shown in the following:

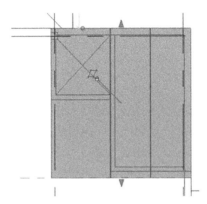

**25.** Go to the 3D view and attach all walls to this newly created roof. For the north outside wall, use the Edit Profile command.

**26.** Go to the Site view. Change the Visual Style to Wireframe.

**27.** Select the new roof from the context tab, click **Vertical** and draw a rectangle to remove the part between the staircase and the elevator as in the following:

**28.** Click (✓) to end the command.

**29.** Load window family M_Skylight-Flat.rfa (Skylight-Flat.rfa).

**30.** Select size = 1200 × 900 (44" × 46") and place it as in the following:

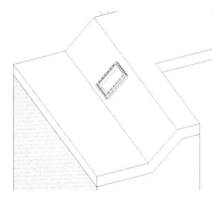

**31.** Go to the 3D view and look at your model from different angles.

**32.** Save and close the file.

## NOTES

CREATING ROOFS • **163**

## CHAPTER REVIEW

1. In Roof by Extrusion, the profile can be open or closed:

   **a.** True

   **b.** False

2. Roof _____ will allow the user to pick walls and specify Overhang and Slope.

3. You can use the Reference Plane command before Roof by Extrusion or inside the command.

   **a.** True

   **b.** False

4. Regarding Shed, Gable, and Hip:

   **a.** You can create them using Roof by Footprint.

   **b.** The default angle for the roof side is 30°.

   **c.** You can define the Overhang value for each side.

   **d.** All of the above

5. Slope Arrow in both Roofs and Floors:

   **a.** The length and angle are very important.

   **b.** You can define slope.

   **c.** Define the level and offset for head and tail.

   **d.** All of the above

6. _____ button will create a hole inside a Roof by Extrusion.

## CHAPTER REVIEW ANSWERS

**1.** b

**3.** a

**5.** d

CHAPTER 9

# COMPONENTS AND CEILING

**This Chapter Contains**

- Placing components
- Creating a ceiling using the Automatic method
- Creating a ceiling using sketching
- Creating a ceiling soffit

## INTRODUCTION

- In this chapter, we will discuss two subjects:
  - Placing components
  - Creating ceilings
- In the first part, we will discuss the definition of components and methods to insert them, whether off-line or online.
- In the second part, we will focus on ceilings, types, and different methods for creating them.

## PLACING COMPONENTS

- Components are elements like Casework, Furniture, Electrical Fixtures, Lighting Fixtures, Planting, and so on.
- Components are still BIM elements that hold information, and you can create schedules out of them, so they are not dummy blocks.
- Component files are RFA files like doors and windows, except they will be loaded from a different command.
- There are three ways to place them into your project:
  - Components are pre-loaded into your current project
  - Components can be loaded using RFA files included with Revit
  - Components can be loaded online
- Some components do not need a host, like furniture and trees.
- Other components, like lighting fixtures, need a ceiling or a wall as a host.
- For each family, you will see different options.
- To start the Component command, go to **Architecture** tab, locate **Build** panel, click the pop-up list of **Component**, click **Place a Component** button:

- Using Properties, you will see the following:

- Those are the pre-loaded component families.
- For some components, you can use the space bar to rotate them.
- To load component family, do the following steps:
    - Go to **Insert** tab, locate **Load from Library** panel, and click **Load Autodesk Family** button:

- You will see the following:

- From the left pane, click Doors (or Windows) you will see the following:

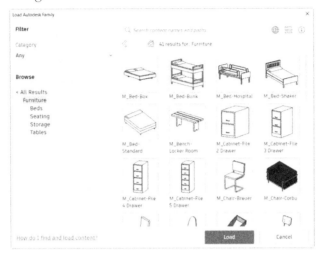

- In the above examples, it will show all furniture in the Autodesk library. You can click the sub-category results click the names under Furniture.
- At the top right, click the shape of the globe to change language and the region of the content you would like to use:

- You can search by typing the phrase to filter your search

- Once you load the new component family, go to **Architecture** tab, locate **Build** panel, click the popup-list of **Component**, click **Place a Component** button:
- To download component files from the web, use any search engine, such as Google, Yahoo, Bing, and so on to find Revit components
- Simply type the name of the desired component with words like Revit Architecture and RFA
- The search engine will list many websites which contain RFA files to download onto your computer.
- Autodesk recommends *https://new.bimobject.com/* as the best website to find what you need. They have more than 37,000 product family, and more than 260,000 parametric BIM object

## EXERCISE 9-1  PLACING COMPONENTS

**1.** Start Revit 2023.

**2.** Open the file **Exercise 9-1.rvt**.

**3.** Go to 00 Ground floor plan.

**4.** Zoom to the right office between grid lines D & E.

**5.** Using the Place Component command, add M_Desk 1830 × 915mm (Desk 72" × 36") at the north side of the office.

**6.** Load the following files from Furniture\Seating:

    **a.** M_Chair-Corbu.rfa (Chair-Corbu.rfa)

    **b .** M_Chair-Executive.rfa (Chair-Executive.rfa)

    **c.** M_Chair-Task (Arms).rfa (Chair-Task (Arms).rfa)

    **d.** M_Sofa-Pensi.rfa (Sofa-Pensi.rfa)

**7.** Load the following files from Furniture\Storage: M_Cabinet-File 4 Drawer.rfa (Cabinet-File 4 Drawer.rfa)

**8.** Load the following files from Furniture\Table:

    **a.** M_Table-Coffee.rfa (Table-Coffee.rfa)

    **b.** M_Table-Dining Round w Chairs.rfa (Table-Dining Round w Chairs. rfa)

**9.** Load the following files from Lighting\Architecture\Internal: M_Floor Lamp - Standup.rfa (Floor Lamp - Standup.rfa)

**10.** Locate the components as shown in the following:

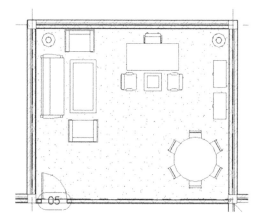

**11.** Do the same for the left office.

**12.** Using the Plumbing folder, load, and place components to create the toilets for men and women.

**NOTE**  *For the men's bathroom, use urinals beside the standard toilets.*

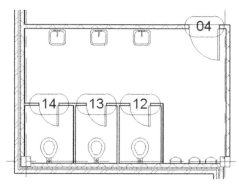

**13.** If you have time, add casework for the cafeteria.

**14.** Save and close the file.

## CREATING A CEILING – INTRODUCTION

- For the first time, we will work on the Ceiling Plans view.
- There are two methods to create a ceiling:
  - **Automatic Ceiling**, where you will move your mouse pointer to a closed area bounded by walls
  - **Sketch Ceiling**, where you will sketch the shape of the ceiling using drawing tools
- We will introduce Ceiling Soffit and some components that are usually used with ceilings.

## CREATING A CEILING – AUTOMATIC

- As the first step, go to the desired ceiling plan view (lots of people forget to do this important step).
- Go to **Architecture** tab, locate **Build** panel, and click **Ceiling** button:

- The Context tab will appear with the Automatic Ceiling mode enabled:

- At the Properties palette, you will see the following:

- Select the desired ceiling family.
- Move your mouse pointer to an area in the building with bounding walls, and you will see a thick red line surrounding the area:

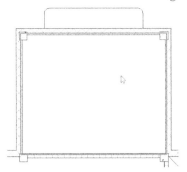

- If you are sure of this area, click once inside it, and you will see the following:

- To select the Ceiling Compound Grid, simply select one of the lines.
- You can rotate and move the whole grid. If you select the grid, the following context tab will appear:

Edit
Boundary

Mode

- Clicking this button means you will start the boundary editing process that was discussed in Walls, Floors, and Roofs.
- Using this option will allow you to set the slope for the ceiling.

## CREATING A CEILING – SKETCH

- If you do not have bounding walls or you want to draw an inventive ceiling, you need to sketch it.
- You will use the normal drawing tools to draw the desired shape.
- Start **Ceiling** command, click the **Sketch Ceiling** button, and you will see the following context tab:

- You can draw any shape you want or select Pick Line or Pick Wall.
- Also, you can create a Reference Plane just like we did with Roofs.
- The boundary line should have the same characteristics of boundaries discussed previously.
- Once you are done click (✓).

**NOTE** *To create an opening in the ceiling for both types:*
- *Use Shaft or Vertical just like we did in Floors*
- *Draw a shape inside the boundary lines; this will be considered as an opening*

- For both types you can add special components in the ceiling, like lights and mechanical equipment (return register, supply diffuser).
- For both types: using **Properties**, set the height of the ceiling:

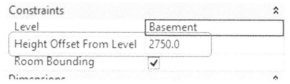

## CREATING A CEILING – CEILING SOFFIT

- A ceiling soffit is the underside of the ceiling, which will connect two ceilings with different heights.
- A soffit is a wall type specially used for this purpose.
- You should control the heights of the soffit carefully to connect the two ceilings correctly.
- Sketch the ceiling first and then add to its edges the needed soffit.

## EXERCISE 9-2   CREATING CEILING

**1.** Start Revit 2023.

**2.** Open the file **Exercise 9-2.rvt**.

**3.** Go to the 00 Ground **ceiling plan**.

**4.** Hide all gridlines.

**5.** Start the Ceiling command and make sure Automatic Ceiling is selected.

**6.** Change the family to Compound Ceiling 600 × 600mm Grid (Compound Ceiling 2' × 2' ACT System) and set height to 2700mm (9'-0").

**7.** Click inside the right office at the north side of the building, then press [Esc] twice to get out of the command. Start the command again; when you click inside the left office, you will notice that Revit took the Cafeteria as well. Accept the selection, and press [Esc] twice to end the command. Select one of ceiling lines from the context tab, select Edit Boundary, and change the boundary to include only the office.

**8.** Zoom to the right office and select one of the ceiling lines. Using the Rotate command, rotate the ceiling grid by 45°. Do the same thing for the left office.

**9.** Notice the two reference planes from the middle of the left vertical wall and the middle of the upper horizontal wall. Use them to sketch the following shape, bearing in the mind the following:

   **a.** The Polygon Circumscribed number of sides = 8, and the radius = 3500mm (12'-0")

   **b.** Circle radius = 2500mm (8'-0")

   **c.** Ceiling Type = Compound Ceiling Plain (Basic Ceiling – Generic)

   **d.** Height offset from level = 2700mm (9'-0")

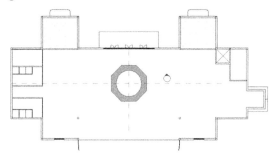

**10.** Draw a soffit wall to cover the circle with height = 300mm (1'-0").

**11.** Create a circular ceiling to cover the shape from the top with height = 3000mm (10'-0").

**12.** Create a camera to see if your work is correct.

**13.** Use Place Component, Load Family, select Lighting\Architectural\Internal, load M_Troffer Light - Parabolic Square.rfa (Troffer Light - 2x2 Parabolic.rfa), and add four lights for each ceiling of the two offices.

**14.** Using the *BIMOBJECT* website, download one of the chandeliers available and place it in the middle of the ceiling you just created.

**15.** Save and close the file.

## NOTES

## CHAPTER REVIEW

1. BIMOBJECT website is the only source for free Revit families:

    **a.** True

    **b.** False

2. The file extension for component family files is _____

3. Soffit is a wall type:

    **a.** True

    **b.** False

4. To create an opening in a ceiling:

    **a.** Draw an inner shape inside the boundary

    **b.** Use the Shaft command

    **c.** Use the Vertical command

    **d.** All of the above

5. To create a ceiling, you can use:

    **a.** Automatic Ceiling

    **b.** Compound Ceiling Grid

    **c.** Sketching the ceiling boundary

    **d.** A & C

6. Use _____ to rotate some of the components.

## CHAPTER REVIEW ANSWERS

1. b

3. a

5. d

## CHAPTER 10

# CREATING STAIRS, RAMPS, AND RAILINGS

## This Chapter Contains

- Creating and editing stairs
- Creating ramps
- Creating railings

## INTRODUCTION

- In this chapter, we will discuss how to create and modify Stairs and Ramps, and of course we have to tackle railings because they work with both of them. So, let's start.

## CREATING STAIRS – FIRST LOOK

- As discussed before, Revit elements are all intelligent elements, and stairs and ramps are no exception.
- You can deal with stairs using one of the following methods:
  - **Standard stairs**: where you will specify the start and end of stair flights, and Revit will take care of the rest, using stair families
  - **Editing standard stairs**: where you will change the shape(s) of the flights and the landing of an existing standard stair
  - **Sketching stairs**: where you can sketch your stairs in any shape

## CREATING STAIRS – STANDARD STAIRS

- Stair families are system families.
- In the standard stairs method, you will specify the start and end of stair flights using a stair family.
- You will use reference planes as a helping tool.
- The final result will depend heavily on the stair family used, so you should select the desired family as your first step.
- To start the Stair command, go to **Architecture** tab, locate **Circulation** panel, and then select **Stair** button:

- The first step is to look at **Properties** palette to select the desired stair family; you will see the following:

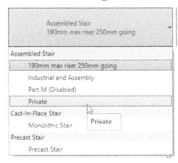

- Why do we do this step first? Because family settings will influence other settings, so selecting the wrong family first, and then changing the family later, may lead to wrong results.
- The context tab will look similar to the following:

- Make sure the **Run** button is on. Select one of five possible shapes:
  - Straight (default option)

- Full – Step Spiral
- Center – Ends Spiral
- L-Shape Winder
- U-Shape Winder

▪ The sixth is not a shape, but it is a mode. The sixth button is **Create Sketch**.

▪ If you select it, the context tab will change to:

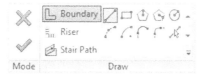

▪ At the right of context tab is the **Railing** button (we will discuss it shortly).

▪ You can create reference planes within the command, or you can create them beforehand (we prefer to create the reference planes within the command, because you will not see them in other floors).

▪ Properties palette will appear as the following (it may change a little bit depends on the family selected):

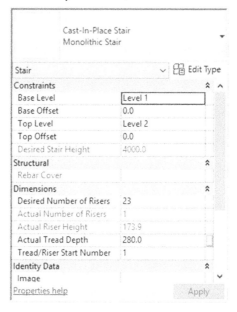

▪ Before we discuss the contents of this palette, let us lay down a simple fact. Revit will create a stair from the current level to the level above. Knowing this, let's discuss the previous image:

- Specify the Base level, and the top level with offsets or without.
- The Desired Stair Height (read only) is the height between the base level and top level.
- Specify the desired number of risers (actual number of risers and actual riser height are read-only fields). Most likely this value is calculated based on two factors: your selected stair family and the height between the two levels.
- Specify actual tread depth.
- Specify tread / riser start number.

■ The Options bar will show the following:

- Specify the stair Location Line. This will decide how you will specify the start and end points of flights and landings; you have several choices to pick from as listed above, with or without offset.
- Specify the Actual Run width.
- Specify whether you want to add an automatic landing between flights, or not.

■ After you control all of these things, specify the points. Do the following:
- Start the first point (the first point will be the first riser of the stairs).
- You will see one flight, representing the whole height of the stairs. At the bottom you will see how many risers were created and how many are left.

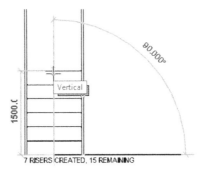

- In the above example, seven risers were created and fifteen remain.
- Whenever you want to stop the flight, click a single click to end it.
- Move up, right, left, and click to start the second flight; the distance between the two clicks will be considered as a landing.

- When you are done, you will see the following:

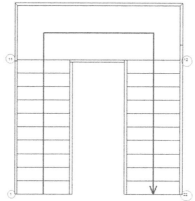

- Notice the numbers at start and end of each flight; they are the number associated with each riser (this stair was drawn from left to right).
- Click the **Railing** button at the right and you will see the following dialog box:

- Select the desired railing family, and then specify the position of the railing, whether on Treads or Stringers.
- Once you are done, click (✓) to end the creation of the standard stair. Below is the final shape:

- You will be able to edit each flight and each landing separately

## CREATING STAIRS – SKETCHING STAIRS

- Start Stair command, from context tabs, select **Create Sketch** button:

- You will see the following context tab:

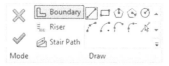

- The **Boundary** and **Riser** buttons will allow you to sketch any shape you want. Boundary is green and Riser is black
- Check the following example, where we sketch two inclined boundaries, with multiple arcs representing the risers:

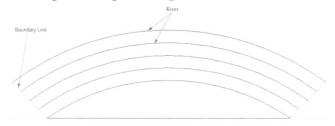

- Once you are done, click (✓) to end the creation of the staircase. Revit will take you back to the previous mode of staircase command

## CREATING STAIRS – CONNECT LEVELS

- If you have typical floors and you want the stiar run to repeat it self multiple times to higher levels or lower levels, then you should use Connect Levels
- When you start Stair command, you will see the following context tab:

- Create a stair from Level 1 to Level 2 (for example)
- Go to one of the four Elevation views
- Under Multistory Stairs context panel, click Connect Levels button
- Select the desired levels (hold [Ctrl] to select multiple levels
- Click (✓) to end the command
- You will see the stair extended to the levels you selected

## EDITING STAIRS

- If you have a stair already in place, simply click it (avoid clicking the railing, or clicking one flight, make sure to select all the stair flights—Use [Tab] key to select it), and at the context tab, you will see the following button:

NOTE *Select Levels button will appear turned off, if you already selected levels for multistory stairs*

- Click the **Edit Stairs** button. The context tab will change to the following:

- You will see the following:

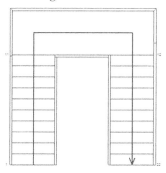

- Each part of the stair can be selected separately:

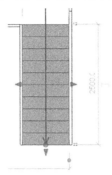

- Change the width of the flight by moving the left or right arrows.
- The bottom arrow will change the number of risers from one flight to the other. Pulling it downward means more risers in this flight, and vice versa.
- Click the landing and you will see the following:

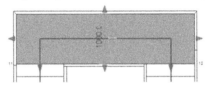

- You can change the dimension of all the landings' sides. To change the flight or landing shapes, you need to convert them first to Sketch Edit capability.
- To do that, and while you are in the stair edit mode, select either flight or landing, and at the context tab, click the **Convert** button.
- You will see the following message:

- This message is telling you that the conversion process is irreversible; click **Close** after reading it.
- Now the Edit Sketch button in the context tab will be enabled and you can customize the shape.

## EDITING SKETCHED STAIRS

- If you have a sketched stair already in place, simply click it (avoid clicking the railing, or clicking one flight, make sure to select all the stair flights—Use [Tab] key to select it), and at the context tab, you will see the following button:

**NOTE** *Select Levels button will appear as turned off if you already selected levels for multi-story stairs.*

- Click the **Edit Stairs** button. The context tab will change to the following:

- Select the desired flight, **Edit Sketch** button will be enabled. You will see three different colors:
  - Green means Boundary
  - Blue means Run
  - Black means Risers

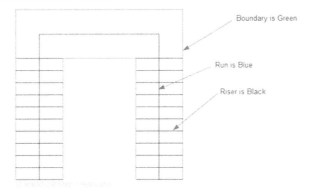

- The context tab will change to the following:

- To change a boundary, select **Boundary** button first. This applies to Risers as well.
- You can edit the shape of a boundary for both the landing and the flights.

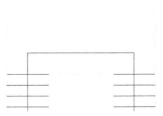

- You can change the riser shapes as well:

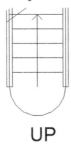

UP

## CREATING RAMPS

- Creating ramps is very similar to creating stairs.
- Ramps are available in a variety of shapes (straight, arc, circular, etc.). They normally cover shallow vertical distances.
- To start the command, go to **Architecture** tab, locate **Circulation** panel, and then click **Ramp** button:

- Using Properties, select the desired ramp family. Set the different parameters (which were discussed in creating stairs) like:

- The context tab will appear as the following:

- This is like Sketched Stair, so all things discussed there are valid here.
- Use **Run** to draw a standard ramp, or use **Boundary** and **Riser** to sketch.
- There are three important points you should know about ramps, and all of them are found in the **Edit Type** button (found in the Properties palette):

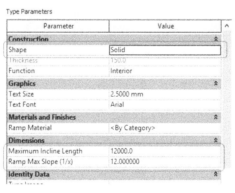

- They are:
  - The Maximum Inclined Length, which is the maximum the ramp run can go without a landing (default value is 12m)
  - Max Ramp Slope (1/x), which is the maximum slope of a ramp

- Shape (thick or solid); check the illustration below:

# CREATING RAILINGS

- By default the railing will be added when you create a stair or ramp.
- Though they are created together, you can modify the railing separately.
- You can add railings for other elements like floors to create balconies.
- Railings have their own families.
- There are multiple methods for dealing with railings:
  - Modify an existing railing of a stair or ramp
  - Create a new railing

### Modify Existing Railing

- To modify an existing railing, do the following:
- Hover over the railing until you see a tool tip as shown in the following (you can use the [Tab] key as we did to select other elements):

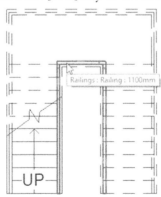

- Click to select. The context tab will change to the following:

- You can **Edit** the **Path** of the railing (the railing path can be opened, but not overlapped).
- You can **Pick** a **New Host**.
- You can **Reset a Railing**, hence removing any type changes applied to this railing.

### Create New Railing

- The first method to create a railing is Sketch a Path:
  - To select this command, go to the **Architecture** tab, click the **Circulation** panel, click the list, and select the **Sketch Path** button:

  - Select the railing family.
  - The context tab will appear as the following:

  - Normal Draw tools are included, along with reference plane tools.
  - You can choose to pick a suitable host (like floors) or Edit Joins.
  - Once you are done, click (✓).
- The second method is Place on Host.
  - To select this command, go to the **Architecture** tab, click the **Circulation** panel, click the list, and select the **Place on Host** button:

- Use this when you want to delete an existing railing on stair or ramp and insert a new one. You will see the following context tab:

- Before selecting the desired stair or ramp, select whether you want to place the railing on Treads or Stringer, then select the desired element.

## EXERCISE 10-1    CREATING STAIRS, RAMPS, AND RAILINGS

**1.** Start Revit 2023.

**2.** Open the file **Exercise 10-1.rvt**.

**3.** Go to 00 Ground floor plan and zoom to the stairwell at the right.

**4.** Hide gridlines.

**5.** Start Stair command. Create the following reference planes:

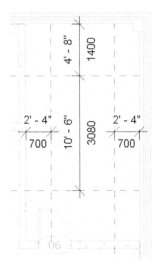

**6.** Create a stair from ***left to right***, using the following information:

    **a.** Family = Cast-In-Place Stair Monolithic Stair Special

    **b.** Desired Number of Risers = 30

    **c.** From Ground Floor to First Floor

    **d.** Actual Run Width = 1400mm (4'-8")

    **e.** Railing = 900mm Pipe (Handrail Pipe) on Treads

**7.** If the warning "Rail is not continuous" appears, close it without doing anything.

**8.** You should receive the following result:

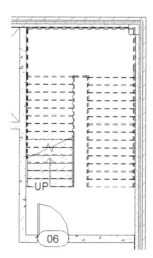

**9.** Select the outside railing, and change the family to Handrail (Guardrail – Pipe Special).

**10.** Change the path of the railing so it will go around the column at the top right.

**11.** Go to the 01 First floor plan, and hide gridlines (if they are not already hidden)

**12.** Zoom to the stairwell.

**13.** Start Stair command. Using reference planes, create the following:

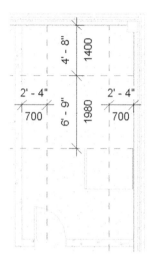

**14.** Create a stair using the same settings, but this time Number of Desired Risers = 20

**15.** Go to North elevation view. Hide the wall covering the staircase

**16.** Select the stair you just created. From the context tab, click Select Level. Holding [Ctrl] select 03 Third level, up to 06 Roof level

**17.** Change the outside railing to be Handrail (Guardrail – Pipe Special), and change the path to avoid the column at the top right

**18.** Use the Floor Edit Boundary command to change the shape of the floor to accommodate all stairs (you can use Shaft command as well).

**19.** While you are at the 01 First floor plan, draw a small railing connecting the railing from the 00 Ground to 01 First floor plans, selecting the floor as your host.

**20.** To make sure you did this correctly, go to 3D view, and go to the wall covering the new stair; hide this wall.

**21.** Go to the Street 1 plan. Hide all gridlines. Zoom to the North entrance.

**22.** Start Stair command, select **Create Sketch** mode, and sketch the following stair:

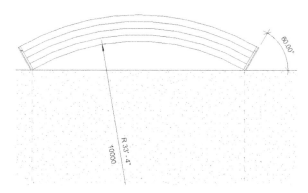

**23.** Take care of the following points below:

    **a.** Family = 190mm max riser 250mm going (7″ max riser 11″ tread)

    **b.** Draw the first riser to be flush with the edge of the floor using a horizontal line, then draw the five arcs

    **c.** From Street 1 to Ground Floor

    **d.** The distance between arcs is 300mm (1′-0″)

    **e.** Railing = Glass Panel - Bottom Fill

    **f.** Since the first step drawn is considered by Revit to be the lowest, check the stair in 3D after finishing and *flip* it

**24.** If you view at it in 3D, you should receive the following result:

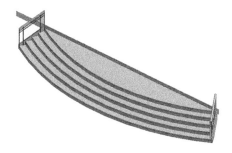

**25.** Using the Railing Sketch Path command (select the slab to be the host of the railing), sketch at the right of the stairs you have just created as in the following sketch:

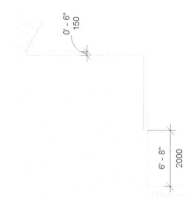

**26.** Use the same Railing family = Glass Panel - Bottom Fill.

**27.** Do the same for the left part of the stair.

**28.** Zoom to the right part of the entrance, where we will create a ramp for the disabled and elderly people. The ramp should cover the vertical distance between the street 1 and 00 Ground floor.

**29.** Go to the Street 1 floor plan.

**30.** Start the Ramp command, select Ramp 1 from Properties, and change the Width to 2000mm (6'-8") and Shape = Solid.

**31.** Using the Reference Plane, create three reference planes as shown in the following:

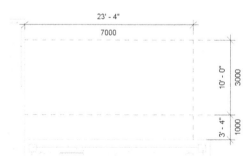

**32.** Make sure that Base level = Street 1 and Top level = Ground Floor.

**33.** Create the following ramp and flip it:

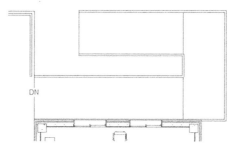

**34.** Do the same to the left part.

**35.** If a warning appears; ignore it.

**36.** Go to the North Elevation view and 3D view to check your work.

**37.** We want to create a stair for the East entrance. Go to the Street 2 floor plan, use Underlay to see the 00 Ground floor plan, and zoom to the east entrance. Start the Stair command, and use the following information:

   **a.** Family = 190mm max riser 250mm going (7″ max riser 11″ tread)

   **b.** Base level = Street 2, Top level = Ground Floor

   **c.** Railing = 900mm Pipe (Handrail – Pipe)

   **d.** Actual Run Width = 3000mm (10′-0″)

**38.** Create the following stair and flip it.

**39.** Go to 00 Ground floor plan, add a railing, 150mm (0′-6″) from the edge of the slab, similar to the following:

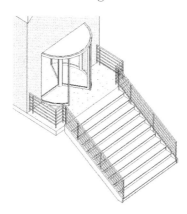

**40.** Go to 00 Ground floor plan.

**41.** Select the stair and click Edit Stairs button.

**42.** Select the stair run (this means select the steps), click **Convert** button.

**43.** When the message appears, read it, and then click Close.

**44.** Click **Edit Sketch** button.

**45.** Change the last three steps to resemble the following (you may use your imagination and create other step shapes). Ensure that the line in the middle (Stair path) reach to the last step:

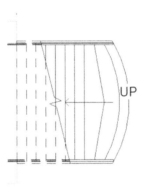

**46.** Go to the East elevation view and 3D view to check your work.

**47.** Go to First Floor plan view and create a railing for the edge of the slab, using a path of 150mm (0'-6") off the edge, and using family = Glass Panel – Bottom Fill.

**48.** Copy it to the 2nd, 3rd, 4th, and 5th floors.

**49.** (Optional) Draw a railing on all balconies using railing family = 900mm Pipe (Handrail – Pipe).

**50.** Look at your model in 3D view.

**51.** Save and close the file.

# NOTES

## CHAPTER REVIEW

1. By default, the maximum Inclined Length a ramp run can go without a landing is 12m (30'):

    **a.** True

    **b.** False

2. _____ and _____ are two types of stairs.

3. In Stair by Sketch:

    **a.** You can draw Boundary lines

    **b.** You can control each flight separately

    **c.** You can sketch the shape of one step

    **d.** Both a & c

4. You should control the inside and outside railing of a staircase together:

    **a.** True

    **b.** False

5. The first step you specify is always the lowest step:

    **a.** Yes, all the time

    **b.** Yes, sometimes

    **c.** No

    **d.** None of the above

6. A railing can be placed on _____ or _____.

## CHAPTER REVIEW ANSWERS

**1.** a

**3.** d

**5.** a

## CHAPTER 11

# CREATING AND MANIPULATING VIEWS

## This Chapter Contains

- Duplicating views
- Creating callouts
- Controlling visibility graphics
- Creating elevations and sections
- View range
- Depth Cueing

## INTRODUCTION

- In this part, we will learn how to:
  - Create a duplicate
  - Create a callout
  - Control Visibility/Graphics
  - Create a section (two types)
  - Create an elevation (two types)
  - Control the view range

## CREATING DUPLICATES

- You can create a duplicate of any view in Revit.
- You can create duplicates of your floor plans and devote them to showing dimensions only (we will do that using visibility control).
- Another reason for duplicating is to show a portion of your model in a certain view using crop region and annotation crop.
- For duplicating 3D views, you can show the 3D presentation for each floor.
- To duplicate a view, do the following:
  - Go to the Project Browser.
  - Locate the desired view.
  - Right-click and you will see the following menu; select **Duplicate View**:

  - There are three types of duplicating; Duplicate, Duplicate with Detailing, and Duplicate as a Dependent. The following is discussion of each one:

### Duplicate

- This method will duplicate only model elements like walls, doors, windows, furniture, and so on but *will not* duplicate anything in **Annotate** tab, **Detail** panel, along with **tags**, **dimensions**, and **text**.
- Any addition of new elements in the source view will appear in the copied view after copying, and vice versa.

- Any addition of new detail elements, tags, and annotation in the source will not appear in the copied view after copying and vice versa, because detail elements, tags, and annotation are view-specific elements.

## Duplicate with Detailing

- This method will duplicate model elements like walls, doors, windows, furniture, and so on along with anything in the **Annotate** tab, **Detail** panel, **tags**, **dimensions**, and **text** (annotation).
- Any addition of new elements in the source view will appear in the copied view after copying, and vice versa.
- Any addition of new detail elements, tags, and annotation in the source will not appear in the copied view after copying and vice versa, because detail elements, tags, annotation are view-specific elements.

## Duplicate as a Dependent

- This method is perfectly suited for big projects which can be cut into zones or areas.
- Duplicate as Dependent means any change in one of the views of any sort will reflect on the others. Views should be identical in all aspects.
- The Duplicate view will appear as a nested view belonging to the parent view, as shown in the following:

## Renaming View

- Once a new view is created, Revit will give it a temporary name.
- You should change the temporary name to a more meaningful name.
- To rename a view, try the following:
  - Go to the Project Browser.
  - Click the desired view name.

- Right-click and select **Rename**:

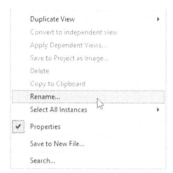

- In the Project Browser, type the new name and press [Enter]:

**NOTE** *Another way is to press [F2], or click the view name twice to rename it.*

### Crop RegionCrop

- Region is a rectangular region shown around the model, and it can be resized to show a portion of the model.
- Go to the desired view, using Status bar, click **Show Crop Region** button:

- You will see the following around the model, click it to show the controls:

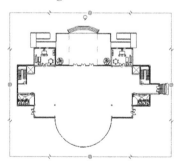

- Using the blue filled circles, you can resize the crop region.
- To turn it off, use the same button in the Status bar.

**Annotation Crop**

- This is created specifically for annotation elements like text and tags.
- Annotation crop can be as small as the crop region or bigger.
- It will hide any annotation element covered or crossed by the crop region.
- The Crop Region should be activated before the Annotation crop is activated.
- To activate it go to Properties and select the checkbox of Annotation Crop as shown in the following:

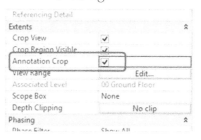

- To see it, click the Crop Region.
- You will see the following:

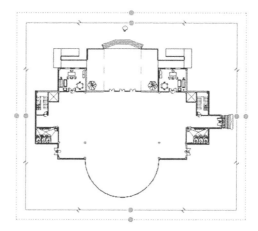

## CREATING CALLOUTS

- Callout is a portion of the model (plan, section, elevation) selected by you to show greater details with larger scale.
- There are two types of callouts in Revit:

- Rectangular shape
- Sketch, any non-rectangular shape
- You have to be in the right view before you issue the command.
- To issue this command, go to **View** tab, locate **Create** panel, and click the popup list of **Callout** button to see the two available options:

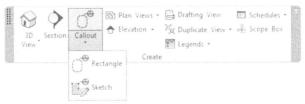

### Rectangular Shape

- For a rectangular shape, specify two opposite corners.
- A new view will be created in the same category of views, holding the name of the current view plus the word Callout. For example, assume your current view name is 00 Ground (floor plan); the temporary callout name will be 00 Ground – Callout 1. Of course, you can rename it.
- Go to the callout view to find out the scale of the view and change it before you start adding any annotation.
- If you click the callout, you will see the following:

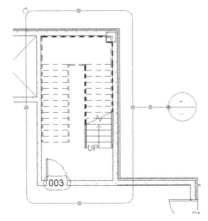

- The blue filled circles allow you to resize the callout and change the placement of the callout head (the callout head will be filled with data by Revit and not by the user).

- Refer to the following illustration:

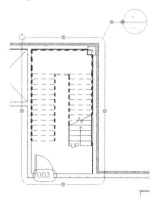

### Sketch

- It is the same as the rectangular callout, except the method of creation is different.
- You will see the following context tab:

- Draw the shape of the callout using Draw tools.
- Once you are done, click (✓). The following is an example of a sketch callout:

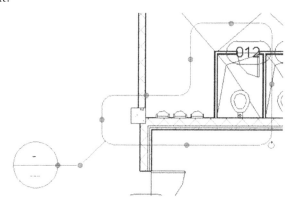

### For Both Types

- If you click either type, the following context tab will be shown:

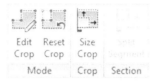

- Using this context tab, you will be able to edit the crop region of the callout, reset the crop region of the callout to its original shape, and resize the crop region of the callout.

## CONTROL VISIBILITY / GRAPHICS

- The Visibility / Graphics dialog box will provide you with full control to show/hide any part of the model or annotation elements.
- For example, select to show the furniture in one view and to hide it in another.
- To issue this command, go to **View** tab, locate **Graphics** panel, and click **Visibility / Graphics** button (type at the keyboard the shortcut VV, or VG):

- The following dialog box will appear:

- In this dialog box, do the following:
  - Click the suitable tab at the top (the most important ones are the first two, Model Categories and Annotation Categories, which will show/hide either model elements or annotation elements).
  - If you do not want to see Furniture elements in a certain view, go to the Model Categories tab and turn off the Furniture and Furniture systems.
  - If you do not want to see Dimension elements in certain view, go to Annotation Categories and turn it off.
  - If you want to click (on/off) for multiple categories in any tab, select them using [Ctrl] at the keyboard. You can use All, None, or Invert (Invert means invert the current selection).
  - In order to speed up the process, you can use **Category name search** to type in the name of the element you want to show/hide, and it will filter the list accordingly, refer to the image below:

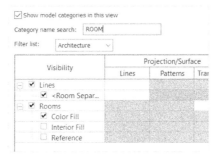

# EXERCISE 11-1  VIEWS — PART 1

**1.** Start Revit 2023.

**2.** Open the file **Exercise 11-1.rvt**.

**3.** Go to 00 Ground floor plan view. Create two duplicates of the Ground Floor plan view and name them 00 Ground – Dimension and 00 Ground – Furniture.

**4.** In 00 Ground – Dimension view, start the Visibility/Graphics dialog box and turn off Furniture, Furniture System, Casework, Lighting Fixtures, and Plumbing Fixtures.

**5.** In 00 Ground – Furniture view, start the Visibility/Graphics dialog box and turn off Dimensions.

**6.** Go to 00 Ground floor plan view. Create a sketched callout for the staircase and elevator and call the new callout 00 Ground – Staircase & Elevator.

**7.** Create a rectangular callout for the two toilets at the left and call it 00 Ground-Toilets.

**8.** Duplicate the South elevation view. Call the new view South-Curtain Wall.

**9.** Show the Crop Region. Resize the Crop Region to show only the curtain wall.

**10.** Hide all levels and Reference Planes. Set the Visual Style = Shaded.

**11.** Go to the 00 Ground floor plan view.

**12.** Save and close the file.

## CREATING ELEVATIONS

- Each template file includes four elevation views.
- You can create as many as you wish. There are two types of Elevations:
  - Building Elevation
  - Interior Elevation
- The creation method for both is the same. Bear in mind that you should be in the plan view to create elevations.
- To issue the command, go to **View** tab, locate **Create** panel, and then click on **Elevation** button:

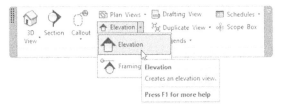

- In Properties, select whether you want Building or Interior:

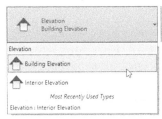

- You will see the following elevation marker:

- This marker is sensitive to walls; it will redirect itself when you get closer to any wall. Once you are satisfied with its location, click to insert it.
- You can control its limits and depth. Click the head (not the circle) to see the following:

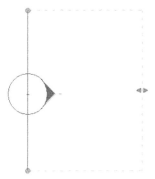

- You can move the two blue dots to resize the coverage size, and the two arrows to resize the depth of the elevation.
- Revit will give the newly created elevation a temporary name; make sure to rename it.
- To go to the elevation view, double-click the head of the elevation marker or the name at the Project Browser.

**NOTE** *To obtain all four sides of the interior room, click the circle (not the head) and you will see four checkboxes; select the side you want to look at. Refer to the following illustration:*

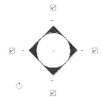

## CREATING SECTIONS

- A section can be created with any angle you want. Even if you create it horizontally or vertically, you can rotate it later. There are two types:
  - Building Section
  - Wall Section
- To issue this command, go to **View** tab, locate **Create** panel, and click **Section** button:

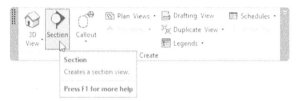

- The Properties palette will show:

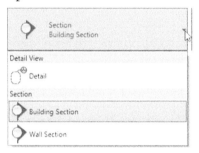

- Select whether you want a Building Section or Wall Section.
- You have to click two clicks to specify the section. The first point will always be the head of the section.
- By specifying the two points, you are specifying the angle of the section.
- After you specify the two points, you will see the following:

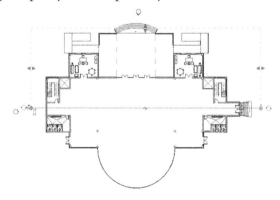

- You can move the two blue dots to resize the coverage size and the two arrows to resize the depth of the elevation.
- Near the head you can see the following:

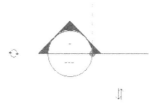

- The two arrows will flip the section head (in the previous example, the section head is pointing north; if you click the arrows, it will point south).
- The two arrows forming a circle are used to cycle the section head shape; you can choose to show Head, Tail, or nothing. This is available at both ends.
- At the middle you will see the following:

- This has no effect on the section functionality. If you want to create a section, and you do not want the section line to appear across the model, you can click this tool. This the result:

- Two new categories in the Project Browser will be created.
- Revit will give all sections temporary names, go to the Project Browser and rename them; you will receive the following:

- If you click the section line, the context tab will change to the following:

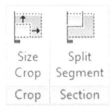

- If you click Size Crop, you will see the following dialog box:

- Using this dialog box, you can resize the section depth and size.
- If you click the Split Segment button, the mouse will change to:

- The Split Segment button will split the section line to appear similar to the following:

**NOTE** *If you have a wall with an angle, and you want to create a section parallel to it, simply use Align command, select the wall, then select the section line.*

- Section will cut into furniture, furniture system, and speciality equipment.

### 3D Section Box

- This only works in 3D views.
- You will see a box covering your model; click it to see arrows at the center of each face to resize the box, hence showing some of the model.

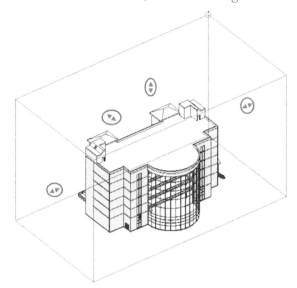

- To activate it, go to any 3D view, and in Properties click **Section Box**:

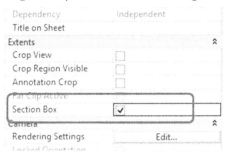

### Orient to View

- This command is another variation of 3D Section Box.
- It can show any saved view in 3D, like floor plans and sections.
- To activate it, try the following:
  - Go to any 3D view.
  - Duplicate it.

- Right-click the ViewCube.
- Select **Orient to View** option.
- Select your desired view:

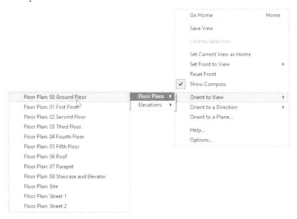

- You will see the following:

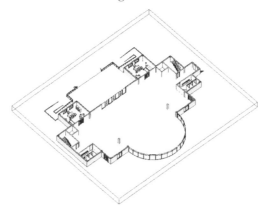

### Selection Box

- Isolate selected elements in the current view or the default 3D view.
- You can select a single element (like stair, door, wall, etc.) or group of elements (like a full room).

- To issue this command, select the desired element or elements in a view, and context **Modify** tab will appear; locate **View** panel and click **Selection Box** button:

*You need to create a duplicate from 3D if you want to keep the selection box result.*

## VIEW RANGE

- To understand the view range for any floor plan (or ceiling plan), look at the following image as your first step:

- Any floor plan is bound by two planes:
  - The Top Plane is the topmost plane and you cannot see beyond it.
  - The Bottom Plane is the bottommost plane and you cannot see below it.
- Each plan has a cut plane which is 1200mm (4-0").
- These three planes are called **Primary Range** of view.

■ To control **View Range**, go to the desired plan:
  • Using **Properties** scroll down, click **View Range** Edit button:

  • You will see the following dialog box:

  • Click **Show** button at the lower left corner, the dialog box will expand and resemble the following:

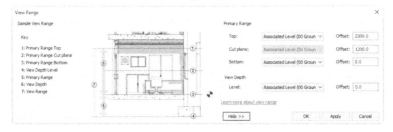

  • For the Top and Bottom planes, there is a Level and an Offset value. You must specify the associated level and then input the offset value.
  • As for Cut plane, you must specify the offset value only.
  • You can specify View Depth, which is below the bottom of the primary range. By default, the Bottom should not be below view depth level, but rather equal or above it. Elements in the View Depth range will be drawn using ***Beyond*** line style.

## PLAN REGION

- Let's assume you want to show two windows with different sill heights on the same floor plan view, bearing in mind you have only one View Range to control.
- The solution is in the Plan Region command.
- You can select a region in your floor plan and define a different View Range. To do that Go to **View** tab, locate **Create** panel, click popup-list of the Plan Views button to show the other buttons, and select **Plan Region** button:

- The context tab will appear as the following:

- The Options bar will include Chain and Offset value.
- The previous Draw panel will allow you to draw any shape you want around the area you want to modify in the View Range. Properties will show the following:

- Click the **View Range** button in Properties and alter the values as you wish.

## EXERCISE 11-2   VIEWS – PART 2

**1.** Start Revit 2023.

**2.** Open the file **Exercise 11-2.rvt**.

**3.** Go to 00 Ground floor plan.

**4.** Create two Building Sections, one that cuts the building vertically, and one that cuts it horizontally (it should be around the middle of the building for each section). The vertical should look to east, and the horizontal should look to north.

**5.** Show head for both ends for both sections.

**6.** Use Gaps in Segments for the vertical section only.

**7.** For the horizontal section, call it Section A-A and for the vertical one, Section B-B.

**8.** Use Split Segment to make Section A-A go through the wall of the staircase (without cutting the staircase itself); check the result of the section before and after.

**9.** Ensure that both sections cover the building and beyond.

**10.** Create the following Building Section and call it Left Office.

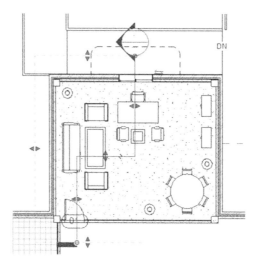

**11.** Go to the Left Office Section view and change the crop region to show only levels 00 Ground and 01 First. Change the View Detail to Medium.

**12.** Create Wall Section as shown below at the east entrance; call it East Entrance:

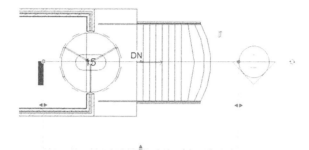

**13.** Go to East Entrance Section view to see the result. Change the crop region to show up to the gable roof and the whole staircase. Extend the levels from the left to go beyond the model.

**14.** Go to the Gentlemen's toilet; create an Interior Elevation inside the toilet showing the four sides. Check each one of these elevations; change the size of the crop region to show the whole toilet floor and ceiling. Set the Visual Style = Shaded.

**15.** Create a duplicate of the 3D view and call it 3D Section B-B.

**16.** Using the Orient to View command, select Section B-B.

**17.** Create a duplicate of the 3D view again and call it 3D East Entrance.

**18.** Using the Orient to View command, select the Section East Entrance.

**19.** Go to 00 Ground floor plan. Zoom to the two toilets; you will notice the two window tags appearing at the left of the wall but without seeing the two windows. This is because the sill height of the two windows is beyond the cut plane. Start the Plan Region command and draw a rectangle around the two windows; click (✓) to finish the command. At Properties, click View Range and set the Cut Plane Offset to 2000 (6-8").

**20.** What happens now after we change the View Range inside the Plan Region?

**21.** Go to the 3D view. Create a duplicate of it and rename it 00 Ground 3D.

**22.** In the newly created 3D view, using the ViewCube, and using the Orient to View command, show the 3D of the Ground Floor.

**23.** Go to the 00 Ground floor plan. Select the stair on the ground floor; using the Modify context tab, locate the View panel and click the Selection Box button. Duplicate the 3D view and rename it "3D Stair." Go to the 3D view and check the Section Box from Properties.

**24.** Save and close the file.

## DEPTH CUEING

- ▪ Depth Cueing is a tool to help you decide to display elements that are farther away from you in a special way using fading.
- ▪ This feature is available in section and elevation views only.
- ▪ As a first step, you need to control Far Clipping:
  - • Far Clip is the depth of the elevation as shown below:

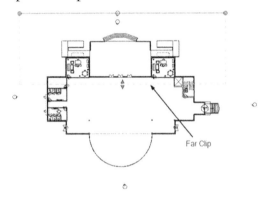

Far Clip

  - • Go to elevation or section view.
  - • From Properties, under Extents, locate **Far Clipping**:

- Click on the button and you will see the following dialog box:

- Select whether you want the clip with a line or without a line (the clipped view will have a frame around it). Click OK.
- Set the Far Clip Offset value.
- To activate the Depth Cueing, try the following steps:
  - Click Visual Style button.
  - Select Graphics Display Options.
  - You will see the following dialog box:

- Expand Depth Cueing and you will see the following:

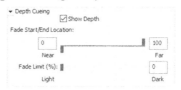

- Turn on the Show Depth checkbox; this will enable the fade of elements.
- Control the Fade Start location (which is labeled Near) and start with value=0. If you increase the value to be more than 0, the fade will

start further from the front view plane, hence, it will reduce the fade of near elements.

- Control the Fade End location (which is labeled Far) and start with value=100. If you decrease the value to be less than 100, the fade limit will be closed to the back view clip plane, hence, it will increase the fade of the far elements.
- When you control the above two sliders, some elements will disappear because Fade Limit % = 0. Increase the value so you can see these elements.
- When done, click OK.

■ To make your view more realsitic, you can turn on the shadows.

## EXERCISE 11-3    DEPTH CUEING

1. Start Revit 2023.

2. Open the file **Exercise 11-3.rvt**.

3. Go to North – Depth Cueing elevation view.

4. From Properties, click Far Clipping button, and set it to the Clip without line option; click OK to end the command.

5. Set Far Clip offset = 32000 (104-0"). You can see more elements were added to the view.

6. Using Visual Style, start the Graphic Display Options command.

7. Expand Depth Cueing.

8. Turn on the Show Depth checkbox. Click Apply to see the effect of fading.

9. Move the Fade End location slider to the left and set it to 50; click Apply, and now you can't see some of the elements.

10. Move the Fade Limit % to 20 and click Apply. Whatever disappeared has reappeared, but with fading color.

11. Move the Fade Start location to 25 and click Apply. You will see the close elements look clearer.

12. Click OK to end the command.

13. Turn on Shadows.

14. Save and close the file.

# NOTES

## CHAPTER REVIEW

**1.** View Range can be controlled for Sections and Elevation views:

    **a.** True

    **b.** False

**2.** _____ and _____ are two types of sections.

**3.** Types of Duplications are:

    **a.** Duplicate

    **b.** Duplicate as Independent

    **c.** Duplicate with Detailing

    **d.** Both a & c

**4.** The Duplicate as Dependent command will create two identical views:

    **a.** True

    **b.** False

**5.** In the Visibility / Graphics dialog box, you can:

    **a.** Show/Hide Model Elements

    **b.** Show/Hide Annotation elements

    **c.** Control the Visual Style of a view

    **d.** Both a & b

**6.** By default, the Cut Plane is at _____ from the level you are in.

## CHAPTER REVIEW ANSWERS

    **1.** b

    **3.** d

    **5.** d

## CHAPTER 12

# ANNOTATION AND LEGENDS

## This Chapter Contains

- Creating and editing dimensions
- Creating and editing text
- Creating legends

## INTRODUCTION

- In this chapter, we will discuss:
  - How to place dimension types on your model
  - How to create Text and Model Text
  - How to create legends

# CREATING DIMENSIONS

- You can create all types of dimensions for your model such as:
  - Aligned (the most used)
  - Linear
  - Angular
  - Radial
  - Diameter
  - Arc Length
  - Spot Elevation
  - Spot Coordinate
  - Spot Slope
- You can create a single dimension or series of dimensions using a single command. Also, you can dimension a whole wall (with or without openings).
- To issue the command, go to **Annotate** tab, and locate **Dimension** panel:

## Aligned

- It can measure both aligned and orthogonal distances.
- Select Aligned. The context tab will repeat the same dimension types, which means if you started any dimension command, you can use another one without the need to exit the current command.
- The Options bar will appear as the following:

- The first part of the Options bar, the reference line, offers the following options:

- Choose between Wall centerlines or faces and Core centerlines or faces.

- As for Pick, you have the following choices:

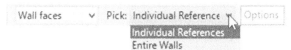

- Individual Reference means you can select any element in the model (grid line, face of wall, etc.), but if you choose the Entire Walls option, the Options button will be enabled; if you click it, you will see the following:

- You can select whether to include openings or not; if yes, choose for them to be measured from widths or centers.
- You can select whether you want to include Intersecting Walls and Intersecting Grids while selecting the entire wall.
- The following is an example of a Wall with an opening:

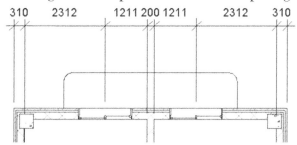

- All of these dimensions are considered to be a single element.

## Linear

- It can measure orthogonal distances only.
- Aligned is more comprehensive.
- Points in elements (like walls) will be highlighted.

- For example:

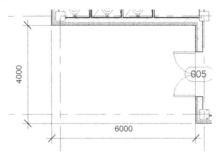

## Angular

- It can measure angles between two reference lines.
- Options bar will show whether your reference lines are Wall centerlines or faces, or Core centerlines or faces.
- For example:

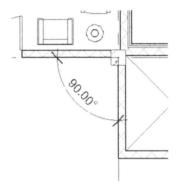

## Radial and Diameter

- It can measure the radius and diameter of an element.
- Options bar will show whether your reference lines are Wall centerlines or faces, or Core centerlines or faces.
- For example:

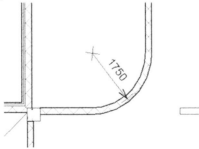

### Arc Length

- It can measure the length of an arc.
- The Options bar will show whether your reference lines are Wall center-lines or faces, or Core centerlines or faces.
- You should select the arc first, then select two intersecting references.
- For example:

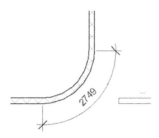

### Spot Elevation

- It can measure the elevation for floors and ceilings.
- The Options bar will appear as the following:

- Select whether you want Leader or Leader with Shoulder:

- For Display Elevation, you have the following choices:

- You can show Actual Elevation, Top Elevation, Bottom Elevation, or both Top & Bottom Elevations (very important in case of elevation or section of floor or roof). For example:

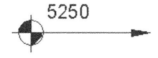

### Spot Slope

- It can measure the slope of an element.
- It can be inserted in any view even in 3D view.
- The Options bar will appear as the following:

- Specify whether you want to show Arrow or Triangle:

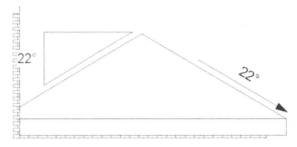

- Specify the offset distance from the reference (should not be less than 1.5mm).

## MEASURE COMMAND

- If you want to measure an element length or between two points without inserting a dimension you can use the Measure command.
- This command can be used in 2D and 3D views
- To issue this command, go to **Modify** tab, locate **Measure** panel, then click one of the two options button (the same command can be found in Quick Access Toolbar):
  - Measure Between Two References
  - Measure Along an Element

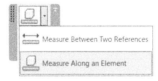

- For the first option (which you can use in 2D and 3D views) you need to select two points in your model and Revit will show you a dimension (which you cannot select or control) and an angle. Using the Options bar, click the Chain option on, and it will allow you to measure the distance between several points, and the total will appear at the Options bar:

- For 3D views, you can use [Ctrl] in order to get Perpendicular distance measurement
- For the second option (only available in 2D), you will click an element like wall, and it will display the dimension (you cannot control from which point to which point in the wall)

## WRITING TEXT

- Writing text in Revit Architecture is so easy.
- There are two types of text in Revit Architecture:
  - Annotation text
  - Model text

### Annotation Text

- Use this text type to write any notes to your model (do not use it to name rooms or areas; this will be handled by Room definition and room tags).
- To issue this command, go to **Annotate** tab, locate **Text** panel, then click **Text** button:

- The mouse pointer will change to:

- The context tab will appear as the following:

- Set the following:
  - Select whether you want a leader with the text or not, and then select the type of leader (one segment, two segment, or curved).
  - Select the justification of the leader with the text written Top Left, Middle Left, Bottom Left, Top Right, Middle Right, or Bottom Right.
  - Select Justification of Text Left, Center, or Right for horizontal justification, and Top, Middle, or Bottom for vertical justification
  - The two tools of Check Spelling and Find and Replace can be reached from the context tab or from the Text panel shown above.
- Once you specify these things, click using the mouse pointer to specify the first point of your text. A new context tab called Edit Text will appear, which appears as the following:

- Type text and then select any part of it, you can specify any or all of the following:
  - Whether you want the text to be Bold, Italic, or Underlined
  - Whether you want the text to be subscript or superscript
  - Whether you want to convert the small letters to capital letters
  - Select whether you want to create a list using four different methods
  - Based on the above, whether you want to decrease indent or increase indent
  - Also, you can choose to increment or decrement the list values
  - Once done, click the Close button to end adding and editing the text
- For example:

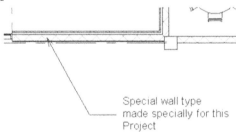

Special wall type
made specially for this
Project

### Annotation Text Family

- You can create your own annotation text type:
- Start the Text command.
- At the Properties click the **Edit Type** button.

- You will see the following dialog box:

- Click the **Duplicate** button to create a new text type and give the new text type a new name.
- Change all or any of the above fields (all of them are self-explanatory).

**Model Text**

- This is text that you will use as a sign for rooms and for the whole building.
- You can give the text a 3D effect by changing the value of the Depth.
- It can be used in all views, but is best used in Sections and Elevation views.
- To start the command, go to the **Architecture** tab, locate the **Model** panel, and click the **Model Text** button:

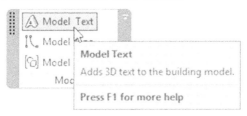

- Once you start the command, you will be asked to select the Work Plane (already discussed).

- Specify the desired work plane and you will see the following dialog box:

- Type the text you want, click **OK**, and then locate the text in the proper place in the elevation or section view. You will receive the following:

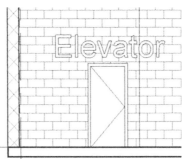

- To change the properties of the text, select it; the Properties palette will appear as the following:

- Change the Text itself, the alignment, material, and depth.

## EDITING DIMENSIONS AND ANNOTATION TEXT

- Clicking an aligned and linear dimension, you will see the following context tab (for other types, only Lable Dimension will appear):

- Edit Witness Lines button will edit the dimensions by changing the location of one of the references or adding a new reference.
- The Properties palette will show the following:

- There are two properties to control:
  - Leader, to add a leader when the text is moved away. For example:

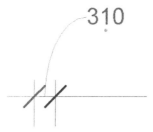

  - The previous illustration shows text moved up the dimension line with an arc leader.
  - Equality Display (discussed previously)

- If you right-click the second blue circle from the end of the extension line of the Aligned dimension, you will see the following menu:

- There are two options in this menu that will help you edit the dimension: **Move Witness Line** (already discussed) and **Delete Witness Line**, which means to discard this line. To understand this: imagine that you used Aligned dimension to dimension from A, to B, and then to C. If you delete the Witness Line of B, the new dimension will measure from A to C.

### Delete Single Aligned Dimension

- Using the [Tab] key you can select one dimension out of a continuous aligned dimension and you can delete it.
- If you click the text in the dimension, you will see the following:

- You cannot put a number in place of the measured distance. But you can:
  - Replace it with text and numbers (if you put only numbers, Revit will reject the editing process)
  - You can put text above, below, prefix, or suffix the measured distance

**NOTE** *You can add Prefix or Suffix in the Dimension Type instead for individual instance.*

- To edit annotation text, try one of the following:
  - To edit the leaders of annotation text, click the desired text once, and the context tab will appear as the following:

  - On the above context tab, you can add / delete leaders (straight or arc shape) of the right and left of the existing text. Also, you can change the location of the leader and the justification of the text.
  - If you double-click the text, the Edit Text context tab will appear to allow you edit the text again.
- If you click the **Check Spelling** button, the following dialog box will appear (which is identical to the one in any word processing software):

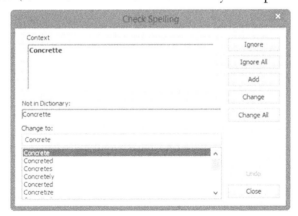

- If you click the **Find / Replace** button, you will see the following dialog box:

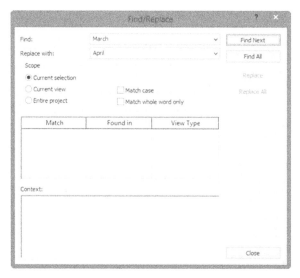

- Type what to find, and with what to be replaced, and then specify whether the desired text exists in the current selection, current view, or entire project.
- Also, click the two checkboxes, Match case and Match whole word only.
- You can use the Find Next, Find All, Replace, and Replace All buttons.
- When done, click the Close button.

## EXERCISE 12-1    DIMENSION AND TEXT

**1.** Start Revit 2023.

**2.** Open the file **Exercise 12-1.rvt**.

**3.** Go to 00 Ground - Dimension view and show grid lines.

**4.** Stretch the bubbles of the top grid lines so it will be easier to add dimensions.

**5.** Using Aligned Dimension command and the default family, create dimensions as follows:

    **a.** Between each vertical gridline, and from A to F

    **b.** Between each horizontal line, and from 1 to 4

**6.** Go to the 01 First – Dimension view and show gridlines.

**7.** Using the Aligned Dimension command from Options bar, select the Entire Wall option. Click Options button and turn on Opening, then Center, and Intersecting Grids. Now, select the wall above the north entrance and locate it above the building.

**8.** Repeat the same process to the left horizontal wall and right horizontal wall, which has the balconies.

**9.** Locate the dimension reading 347 (1' – 1 21/32") at the left of the left wall and move the text away from the dimension. Do the same thing for 347 (1' – 1 21/32") at the right of the right wall. There are two more of the 347 (1' – 1 21/32"); using the Delete Witness line, discard both of them.

**10.** Below shows the final result:

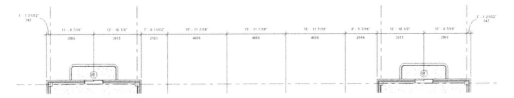

**11.** Put in the following dimensions as shown below (always use wall face, except for the radial dimension):

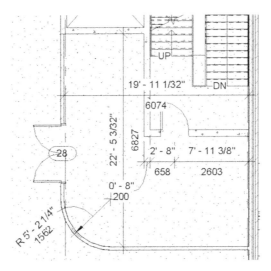

**12.** Put more dimension in the 01 First – Dimension plan.

**13.** Under Ceiling Plans, create a duplicate from 00 Ground and call it 00 Ground - Dimension. Use Spot Elevation (with Leader) as shown below:

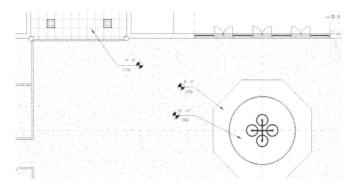

**14.** Under Sections (Building Section), go to the Section A-A view. Zoom to the top, and input the following Spot Slope dimensions:

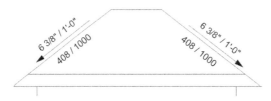

**15.** Under Sections (Building Section), go to Left Office view. Add the text and Spot Elevation as shown below:

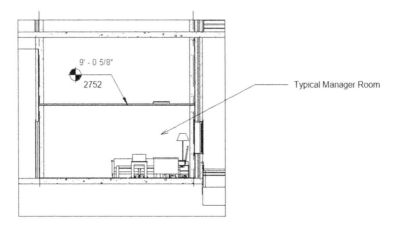

**16.** Under Sections (Wall Section), go to the East Entrance view. Add the text as shown below:

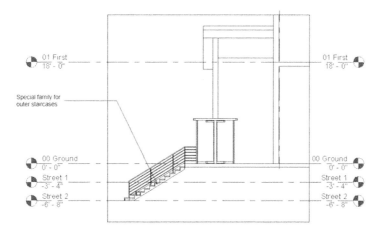

**17.** Go to 00 Ground – Dimension and add Spot Slope for the two flights of the two ramps.

**18.** Go to Sections (Building Section) go to the Section A-A view.

**19.** Zoom to the manager's room door at the fifth floor at the right and start the Model Text command. From the dialog box, select the Pick a plane option and click OK; now pick the wall that contains the door, and when the Edit dialog box comes up, type **Manager Room** and place it beside the door. Press [Esc] a couple of times to get out of the command.

**20.** The text is large. Select it, and at Properties, click the **Edit Type** button, then **Duplicate** and name the new family 200mm Arial. Change the Text size to be 200 (0'-8"), and then click OK.

**21.** While the text is still selected, select the new created family, and then reposition the text as shown:

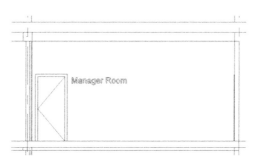

**22.** Using 2D and 3D views try to use the Measure command for Measure Between Two References with and without the Chain option.

**23.** Save and close the file.

## CREATING LEGENDS

- There are two steps to do Legends in Revit:
  - Create a legend view.
  - Insert the desired Symbols and Legend Components. To create similar table-like view, you can use detail line and text.

### How to Create a Legend View?

- A legend view is a separate view not created automatically by Revit.
- To create an empty view, try the following:
  - Go to the **View** tab, locate the **Create** panel, click the **Legends** button to show the list, and select the **Legend** option:

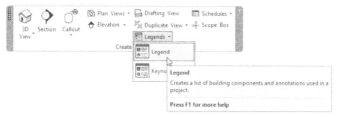

  - The following dialog box will be displayed:

- Type the name of the new view and the view scale, and then click OK.
- If you check the Project Browser, you will see the following:

### Insert Legend Components

- To insert Legend Components, try the following:
- Go to the **Annotate** tab, locate the **Detail** panel, click the **Component** button to show the list, and then select the **Legend Component** option:

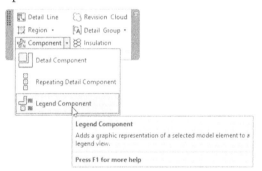

- The Option bar will change to the following:

Family: Doors : M_Double-Flush-Dbl Acting : 1830 x 21  ∨    View: Elevation : Frc ∨    Host Length:  914.4

- Select one of the families you used in your project.
- Select the view, either Floor Plan view or two Elevations (Front and Back).
- Place the component in the view.
- Select another family and so on . . .

### Insert Symbols

- To insert symbols, try the following:

- Go to the **Annotate** tab, locate the **Symbol** panel, and click the **Symbol** button:

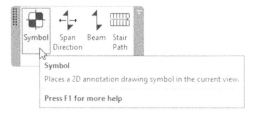

- Depending on the current view, the list of available symbols will change. For example, if you are in Legend view, you will have huge list of symbols to which you can add. But if you are in Floor Plan view (for instance) the list will shrink to two symbols.
- Of course, you can use the **Load Family** button to load the desired symbols from the library.
- To create a table-like list, go to the **Annotate** tab, **Detail** panel, and click the **Detail Line** button:

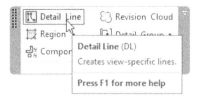

- The context tab will show the following:

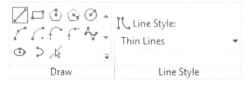

- Using Line Style, select the desired line style:

- Then select the desired drawing tool to draw the shape.
- Use Text as discussed in this chapter to type any clarification for your legend.

**NOTE** *After you insert a legend view to one sheet, you can select it, and Copy it to Clipboard, then Paste it in another sheet(s).*

## EXERCISE 12-2 CREATING LEGENDS

**1.** Start Revit 2023.

**2.** Open the file **Exercise 12-2.rvt**.

**3.** Create a legend view and call it **Door Legend**. We used three Door types with different sizes. The legend should appear as the following (use only the relevant dimensions):

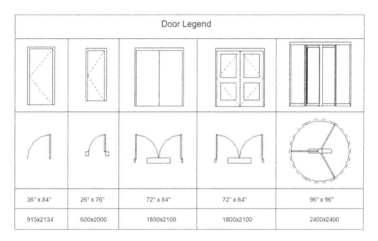

| Door Legend | | | | |
|---|---|---|---|---|
| 36" x 84" | 26" x 76" | 72" x 84" | 72" x 84" | 96" x 96" |
| 915x2134 | 600x2000 | 1800x2100 | 1800x2100 | 2400x2400 |

**4.** Save and close the file.

## NOTES

## CHAPTER REVIEW

**1.** There are two types of text in Revit:

 **a.** True

 **b.** False

**2.** You will create the Legend view from _____ tab.

**3.** Spot Elevation:

 **a.** Can have Leader

 **b.** Can have Shoulder

 **c.** Must have Leader

 **d.** Both a & b

**4.** You should draw the lines representing a table in a legend view:

 **a.** True

 **b.** False

**5.** All of the following are dimension-editing tools in Revit:

 **a.** You can move the dimension text out of the dimension block

 **b.** Move witness line

 **c.** You can change the measured distance

 **d.** Delete witness line

**6.** Using _____ you can write text on walls in sections and elevations.

## CHAPTER REVIEW ANSWERS

**1.** a

**3.** d

**5.** c

## CHAPTER 13

# VISUALIZE IN REVIT AND PRINTING

## This Chapter Contains

- Dealing with Cameras
- Creating Walkthroughs
- Solar Studies
- Basic Rendering
- Rendering Settings
- Working with Lighting
- Enhancing Renderings
- Creating Sheets and Printing

## DEALING WITH CAMERAS

- By default, the 3D view will show orthographic view but you can change it to Perspective either by right-clicking the ViewCube and select Perspective or by creating a camera.
- We already know how to place a camera in a plan but we do not know how to control it.
- You can select and modify elements (not all modifying commands are available) in a perspective view.
- Perspective views can be added to sheets.

### How to Create a Camera

- To create a perspective view, try the following steps:
  - Go to a plan view.
  - Go to **View** tab, locate **Create** panel, expand **3D View**, and click **Camera** button:

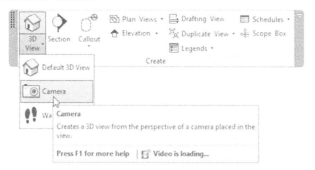

  - Or you can go to Quick Access Toolbar and issue the command as shown below:

  - In the Options Bar, select **Perspective** option and set an **Offset** for the camera **From** one of the levels, as shown below:

  - Select two points; the first one is the camera position, and the second one is the target location.
  - Revit creates a new 3D view and opens it; rename it as needed.
  - To control the extents of the camera, simply select the border of the view and change it.
- The output view can be modified by changing the camera properties using controls or the camera properties.
- To control the camera, try the following:
  - Open a plan view.

- In the Project Browser, right-click on the 3D perspective view (not the plan view) and select Show Camera. The camera displays in the plan view.

- Drag the controls for the camera, target, and far clip plane, as shown:

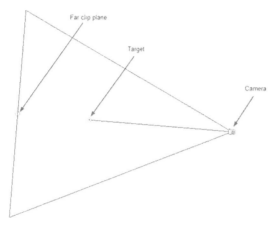

- Control cameras in Properties and scroll down to the Extents and Camera sections, as shown below:

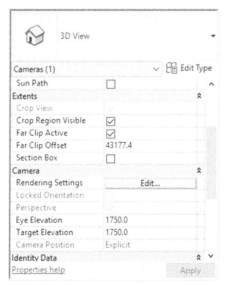

### View Size

- If you select a camera, a context tab will show the Crop panel, which contains the Size Crop button:

- The Crop Region Size dialog box will appear, as shown below; you can set the width and height of the view. Select whether to change the Field of view (without modifying the scale) or change the Scale of the view using locked proportions:

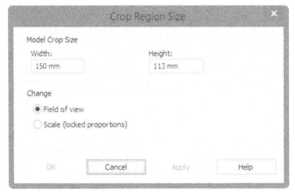

### Reset Target

- While you are in any camera perspective view, you can move the camera using the wheel+[Shift] key. To return to the original camera, click the Reset Cameras button.

**NOTE** *While you are in Camera view, the view is cropped.*

- To make the view uncropped click Do Not Crop View button from View Control bar:

## EXERCISE 13-1    PERSPECTIVES

1. Start Revit 2023.

2. Open the file **Exercise 13-1.rvt**.

3. Go to the 00 Ground plan view.

4. Start the Camera command.

5. Create an exterior view and name it **Curtain Wall** as shown in the following (adjust the crop region as needed to look just like the image):

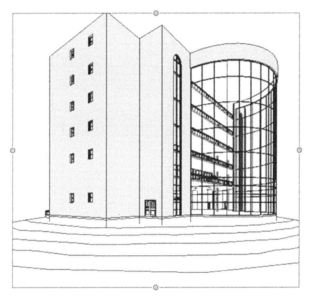

6. Change the visual style to be Shaded.

7. Switch between Perspective and Orthographiccs.

8. Duplicate the Curtain Wall view and name it Curtain Wall from top.

9. Adjust the Eye Elevation to be 30000 (100'-0").

10. Adjust the crop region.

11. Go back to the 00 Ground floor plan.

**12.** Place a new camera in the Manager Office at the top left, similar to the following image:

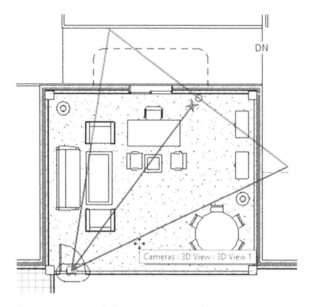

**13.** Name the new view of the Manager Office.

**14.** Click Do Not Crop View from View Control bar to make it uncropped

**15.** Save and close the file.

## CREATING WALKTHROUGHS

- To create a Walkthrough, simply create a path containing numerous cameras, each showing part of the walkthrough movie.
- When you start, you will specify a camera (Revit will show a red dot); this is called a Key Frame.
- You can edit each frame individually.
- Finally, you can export the whole movie to an AVI file.

### How to Create a Walkthrough

- To create a walkthrough, try the following steps:
  - Though you can create a walkthrough in different types of views, normally we will create it in Floor Plan view.

- Go to **View** tab, locate **Create** panel, expand 3D View, and click **Walkthrough** button:

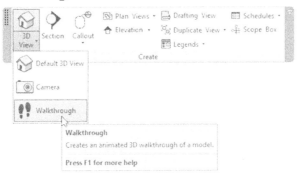

- In the Options Bar, set the perspective and offset from the level as needed, as shown below:

- Click to inset the first camera (key frame).
- Keep on adding key frames as needed. You can change the offset if required. When done, in the context tab, click the Finish Walkthrough button.

### How to View and Modify a Walkthrough

- To view the walkthrough, go to the Project Browser, find the Walkthrough category, locate the walkthrough you created (you can rename it), and then double-click it to view it.
- To edit the walkthrough, click the edge of the view, and the following context tab will be displayed:

- Click the **Edit Walkthrough** button and the following will be displayed:

■ Using the Previous Frame and Next Frame buttons, you will view the walkthrough one frame per click. Using the Previous Key Frame and Next Key Frame you will view the walkthrough one key frame per click.

- To view the whole walkthrough as a movie, use the Play button.
- If you are in a plan view, use the Open Walkthrough button to open the frame which is specified in Option bar.
- While you are in any frame, you can move the camera using the wheel+[Shift] key. To get things back to the original camera, click the Reset Cameras button.
- While you are in a frame, do not click the empty area because Revit will show the following message:

### How to Set Walkthrough Frames

■ You can change the number of frames between key frames.
■ To do this, go to Properties, find **Other** section and click the button next to the Walkthrough Frames parameter, as shown below:

■ You will see the following dialog box and you can do the following:
- Set the Total number of Frames.
- Change Frames per second.

- If you check off the Uniform Speed checkbox, you can change the Accelerator to each key frame.

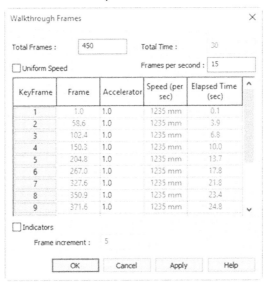

## How to Export a Walkthrough

- To export the walkthrough, try the following steps:
- Open the Walkthrough.
- Using the File Menu, click **Export /Images and Animations / Walkthrough**:

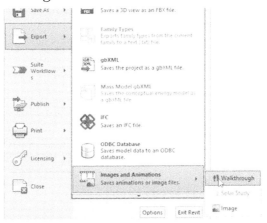

- You will see the following dialog box. Under **Output Length**, specify whether you want all frames or some of them (specify the start and end accordingly). Under **Format** section, specify the Visual Style. Use the current dimensions or specify a zoom scale factor:

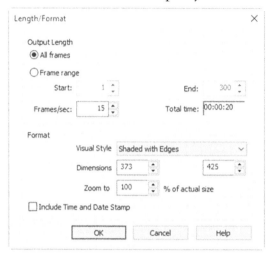

- When done, click OK.
- Save the output file to the desirable destination.
- You will see the following dialog box. Specify the format and quality and click OK.

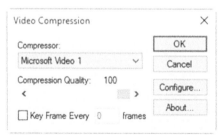

## EXERCISE 13-2   CREATING WALKTHROUGHS

**1.** Start Revit 2023.

**2.** Open the file **Exercise 13-2.rvt**.

**3.** Open the 00 Ground floor plan view.

**4.** Start the Walkthrough command.

**5.** Pick points to create a path that comes in through the north entrance, then goes to the right side into the office, circulating around the furniture. When done, click Finish Walkthrough.

**6.** Using the Project Browser, find the Walkthroughs section, rename the only walkthrough "Into the Office." Double-click it, click the edges of the view, and then click the Edit Walkthrough button.

**7.** In the Options bar, set the Frame to 1, change the Visual Style to be Shaded, and then click the Play button.

**8.** From Properties, under Other, click the Walkthrough Frames button and set the Total Frames to 450. Test the Walkthrough again and you will find it slower this time.

**9.** Using Export, create an AVI file for your walkthrough, leaving everything to default.

**10.** Save and close the file.

## SOLAR STUDIES: INTRODUCTION

- Solar Studies is to switch on the Sun path and the Shadows, to study how your building is affecting the environment and the other neighboring buildings.
- There are three types of solar studies: Still, Single Day, and Multi-Day.

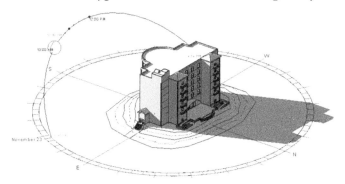

- You can export the solar study as single picture or as an animation.

■ To see the Sun Path and Shadows, turn them on in the View Control Bar, as shown below:

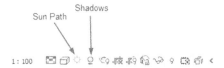

■ As a first step, set the true north of your project.
■ Then manage the location of your project by doing the following:
  • Go to **Manage** tab, locate **Project Location** panel, and click **Location** button:

  • You will see the following dialog box; go to the Location tab, under Define Location by. Set the Default City List, as shown below, and set your city in the section beneath it:

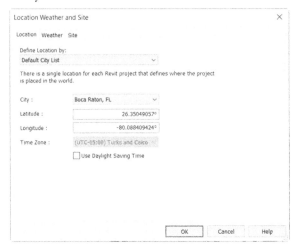

- Or select Internet Mapping Service, as shown below, and type the name of your city:

- You are ready now. Duplicate 3D views and specify the desired type of solar studies.
- Use Graphics Display Options edit options such as the intensity of the sun, the darkness of the shadows, and the background. Do the following:
  - Using the View Control Bar, click the Visual Style button and select Graphics Display Options.

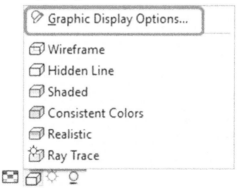

- You will see the following dialog box to control both Shadows and Lighting:

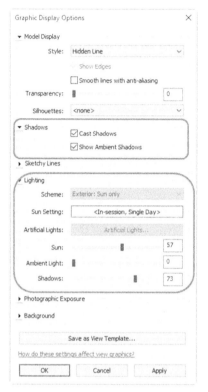

## SOLAR STUDY TYPES

- There are three types of solar studies; still, single day, and multi-day. There are also several preset studies you can use.

### How to Create Still Solar Studies

- To create still solar studies, try the following:
- Using View Control Bar, expand Sun Path drop-down list and pick Sun Settings:

- In Sun Settings dialog box, under the Solar Study section, pick Still.
- Under Settings, specify Location, Date, and Time.
- Or, under Presets section, select an existing study and click Duplicate, type a name for the study, and click OK.

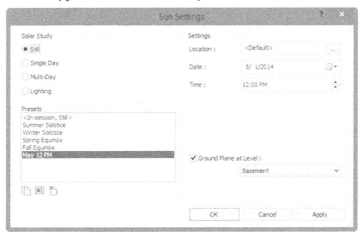

- When done, click OK to close the dialog box and apply the still solar study to the current view.

### How to Create Single Day Solar Studies

- To create a single day solar study, do the following steps:
- Open Sun Settings dialog box.
- Under Solar Study section, pick Single Day option.
- Under Settings section, set Location, Date, and Time (start and end). Or, select the Sunrise to sunset option to automatically calculate the times based on the date. Select Time Interval (15, 30, 45 minutes, or 1 hour). The Frames value is read-only.
- You can use an existing Preset; just Duplicate it and name it.

- Click OK to close the dialog box.

- To view solar study, go to the View Control Bar, and pick Solar Study, as shown below:

- A context tab will appear:

- Control the animation using Preview and Play panel.
- Control the speed, sunlight intensity, and shadow intensity using Display panel.
- Control the Study Type whether Still, Single Day, Multi-Day, or Lighting using Study Type panel.
- Control which preset to use, date and time, and which frame to display using Presets and Data panel.

### How to Create Multi-Day Solar Studies

- To create a multi-day solar study, try the following:
- Open Sun Settings dialog box.
- In Solar Study section, select the Multi-Day option.

- Under Settings section, specify Location, Date, and Time (start and end) and select Time interval (One hour, One day, One week, or One month).
- You can use an existing Preset; just Duplicate it and name it.
- Click OK to close the dialog box.

- Do exactly the steps done for single day to view the animation.

## EXERCISE 13-3    SOLAR STUDIES

**1.** Start Revit 2023.

**2.** Open the file **Exercise 13-3.rvt**.

**3.** Go to **Massing & Site** tab, locate **Conceptual Mass** panel, and click **Show Mass Form and Floors**. You should see masses depicting neighbor buildings.

**4.** Go to the Site floor plan view, click Project Base Point, and set the True North to 20°.

**5.** Duplicate Site floor plan view and rename it Site-True North. Open the new view. In Properties, set the Orientation to True North.

**6.** Start the Location command.

**7.** In Location tab, using Internet Mapping Service map, type Boca Raton and then click Search. Select the Use Daylight Savings Time option, and then click OK.

8. Go to the existing 3D view. Duplicate it and name it 3D Still Solar Study.

9. Set the Visual Style to Hidden Line and turn Shadows On.

10. Switch the Sun Path On and select Continue with the current settings. Select Sun Settings. In the Sun Settings dialog box, Solar Study section, select the Still option. In the Settings section, specify today's date. Test the following times: 9 AM, 10 AM, and 11 AM (set the Ground Plane level at 00 Ground).

11. In the Project Browser, right-click on the 3D Still Solar Study view and Duplicate with Detailing. Rename the new view 3D Single Day Solar Study.

12. Turn off the sun path.

13. Modify the Sun Settings and create a new Single Day solar study from sunrise to sunset, using today's date and a Time Interval of 30 minutes. Click OK, and then Preview Solar Study (set the Ground Plane level at 00 Ground).

14. In the Project Browser, right-click on the Single-Day Solar Study, and Duplicate with Detailing.

15. Rename the new view as 3D Multi Day Solar Study.

16. Modify the Sun Settings and create a new Multi-Day solar study for one year at Noon with a Time Interval of One month.

17. Preview the solar study animation as full, then use the Options bar and Next Frame button to go through each month by itself.

18. Save and close the file.

## RENDERING IN REVIT – INTRODUCTION

- Rendering in Revit is mainly a complementary function rather being fundamental. It can produce still rendered images in high quality, but if you want better images, use the likes of 3ds Max and Maya.
- The Revit rendering engine is the Autodesk Raytracer.
- Prepare your view to be rendered. Set the camera view to create the proper 3D view. Since almost all of the Revit elements have material

defined inside them by default, there is no need to assign materials for the elements. Finally, add lights, components like furniture, trees, people, and other accessories.

## RENDER COMMAND IN REVIT

- To render a view in Revit, try the following:
  - Open a 3D perspective or orthographic view.
  - In the View Control Bar, click **Show Rendering Dialog** button. This option only displays if you are in a 3D view:

- You will see Rendering dialog box as shown below:

## Quality

- Select from the existing list as shown below, or create your own Custom (view specific) settings (Draft is the lowest quality and shortest, and Best is the highest and longest):

## Output Settings

- Send the rendering to the screen or to a printer. If you pick the Printer option, select the DPI (dots per inch). The Width and Height of the view are displayed in pixels and the image size is displayed in megabytes:

## Lighting

- Under the Lighting section, select a Scheme whether to use the sun and/or artificial lights, set up the Sun Settings, and finally set the artificial lights:

## Background

- Under the Background section, set the Style for the sky by selecting from several options of cloud, a color, an image, or Transparent. (If you export the rendered image to PNG or TIFF format, the alpha channel reserves the transparent background to be used by other software packages.):

## Image

- After creating a rendering, Adjust Exposure, Save it to Project (it will be saved as view in the Project Browser), and export the image:

- The Exposure Control dialog box appears as the following. Try each one of the sliders to make the rendered image, highlights, and shadows brighter or darker. Then control the saturation and white point:

- In order for these settings to take effect, wait for the next render.

### Display

- Use the Show the model button to show the model on the screen. Accordingly, Revit will display the model and the button changes to Show the rendering to go back to the rendered image.

### Render and Render Region

- When all of the above settings are controlled, select whether you want to render the view displayed on the screen or you want to render a part (region) of it.
- To do the Region option, click the checkbox and zoom out to see the red boundary displayed; click it and resize it.

## RENDER AND LIGHTS

- Lights in your model are called artificial lights in the Render command.
- Make sure to turn on light sources before any rendering. To do this, perform the following:
  - Type VG or VV to open Visibility/Graphic Overrides dialog box.
  - Using the Model Categories tab, expand the Lighting Fixtures and select the Light Source option:

### Setting Up Light Groups

- For interior renderings, and in order to save rendering time significantly, create light groups just for the room(s) you are working on.
- To create a light group, try the following:
  - Open the ceiling plan view, which contains lights you want to group.
  - Select one of the lights.
  - In the Options Bar, locate the Light Group drop-down list and select an existing group or select Edit/New option to create a new group:

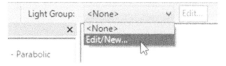

- If you select Edit/New . . . , you will see the following dialog box:

- Click **New** button to create and name a new light group; you will see the following dialog box. Type in the name and click OK:

- Select the other lights, and from Options Bar, select the newly created group; you will see the following warning:

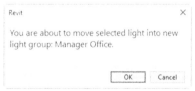

- Click OK to accept.
- To activate the light group, start the Render command, under Lighting, and select a scheme with Artificial Light mentioned. The Artificial Light button will be enabled; click it and turn on your light group and turn off the Ungrouped lights.

**Adding Trees and People**

- To make the scene more vibrant, you can:
- Add trees; Revit is equipped with deciduous and coniferous trees of many types.
- Add people components (found in the Library under Entourage) that are specifically created for rendering. In a plan, the person displays with a point indicating the direction in which they are facing. In a 3D view, they display as outline. In a rendering or Realistic visual style, they will display as they have been created.
- After you insert entourage component, do the following steps:
  - Select it, click Edit Type (in the below example, the female entourage was selected), you will see the following dialog box:

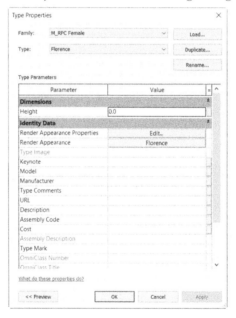

- Click Edit button besides **Render Appearance Properties**, you will see the following:

- If you decide to use Entourage people in a walkthrough, you can control whether or not, you want to Cast Reflection on other surfaces. For Jitter, whether to control the transition between frames in a walkthrough, if you turn it on, it will be smoother. Finally, if you select Billboard to be turned on, that means the entourage will keep looking at the camera, while the camera is moving.
- The above options are only for People, for transportation, you will control the color of the vehicle.

- Click the button besides **Render Appearance**, you will see the following:

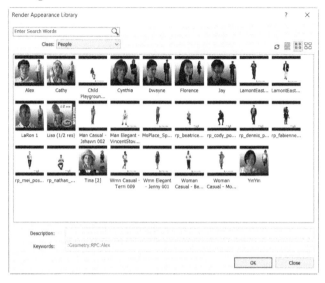

- From Class drop-down, select the desired class, then select the desired shape you want to use, click OK to end the option
- Click Visual Style, and select the first option Graphic Display Options, select Realistic, you will see the following:

- Click RPC drop-down, and select one of the following:
  - Override – All Billboard, which means all the entourage elements will be Billboard will keep looking at the camera, while the camera is moving
  - Use Render Appearance, which means if you render, all entourage will use the original settings
  - Do not Display, entourage will disappear while rendering

## EXERCISE 13-4  RENDERING IN REVIT

1. Start Revit 2023.

2. Open the file **Exercise 13-4.rvt**.

3. Go to Site floor plan, and add M_RPC Tree - Fall.rfa American Beech 6 meters (RPC Tree - Fall.rfa American Beech 20') trees as shown below:

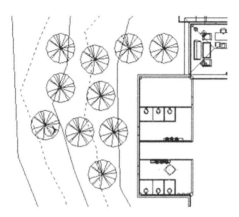

4. Using Entourage folder in the libraries, add M_RPC Beetle (RPC Beetle), and add near right side entrance, set the Offset from Host = -1000 (-3'-4")

5. Go to 3D view and use Realistic visual style

6. Select the car added, click Edit Type, Duplicate, and call the new car: My_Car, from Render Appearance select Class = Automobiles, and select Honda CR-V SUV 2017, click OK. Select Render Appearance Properties Edit, set the Car Paint to Green, click OK twice. Move the car so it will not crash into the wall

7. Go to 00 Ground Ceiling plan and zoom to the top left office. Create a group of lights and call it Manager Office and add the twelve lights to it.

8. Go to 00 Ground Floor plan and add the three Standup lamps to the same group.

9. Using the Entourage folder in the libraries, add M_RPC Female.rfa (RPC Female.rfa), and M_RPC Male.rfa (RPC Male.rfa), and add some females and males to stand around the main desk (reorient them if needed).

10. Use the same technique you used above and change Render Appearance of one of the male entourage.

11. Go to Visibility / Graphics dialog box and turn on the Light Source.

12. Go to Curtain Wall view under 3D Views and switch the Shadows on, render the view using Autodesk Raytracer, Medium, Exterior Sun Only, and Sky: Very Few Clouds.

13. Using Save to Project, save the rendered image to the project under the name Curtain Wall – Exterior.

14. Go to Manager Office view under 3D Views, and switch the Shadows on, render the view using Autodesk Raytracer, Medium, Interior Sun and Artificial, and select only the Light Group named Manager Office.

15. Using Save to Project, save the rendered image to the project under the name of Manager Office Rendered. Export the image to your exercise folder under the same name.

16. Save and close the file.

## CREATING SHEETS

- Construction sheets in Revit are very simple and straightforward.
- You will create a sheet based on a title block template (in this chapter we will use the ones that come with Revit).
- Then add one or more views into the sheet.
- There are two ways to create a new sheet:
  - Go to Project Browser, locate **Sheets (all)**, right-click, and select **New Sheet** option:

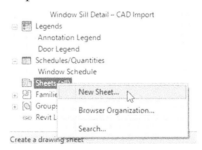

- Go to **View** tab, locate **Sheet Composition** panel, and then click **Sheet** button:

- Using either way, you will see the following dialog box:

- If the title block is listed, select it and click OK, if not, click Cancel, then go to Insert tab, locate Load from Library panel, click Load Autodesk Family, make sure you are seeing the right Region, from the left pane click Titleblocks, select the desired one, and click Load.

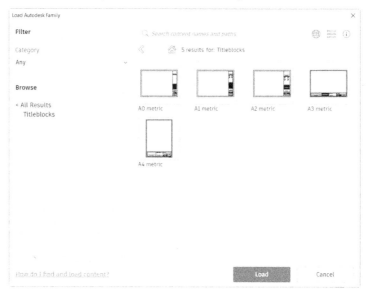

- Create a new sheet and select the loaded titleblock, you will see the following as shown below:

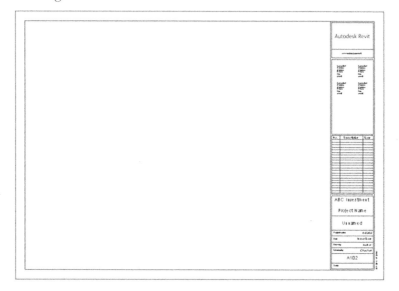

- At the lower right side of the title block, you will see some information already filled in, like the client's name and project number. This information was filled into the Project Information dialog box at the beginning of this book.
- Other pieces of information can be filled either by clicking the field in the title block or by going to the Project Information dialog box and editing it there.
- The following is an example of editing the data inside the title block:

- Another way to change information is the Properties palette:

- At the Project Browser, you can edit the sheet name and number. Click the sheet, then press [F2] or right-click, and select Rename, and you will see the following dialog box:

- Type the new number and name and Revit will reflect that on the sheet.
- Before you start inserting views in the sheets, you can display the Guide Grid.
- Go to **View** tab, locate **Sheet Composition** panel, and click **Guide Grid** button:

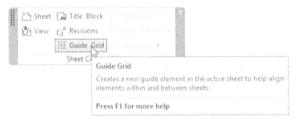

- You will see the following dialog box:

- Under **Create new**, type the name of the new grid, and then click OK.
- You will see a grid over the paper. Click one of its edges and resize it.
- At the Properties palette, you will see the following:

- Edit the Guide Spacing to change the value.

- You will see the following:

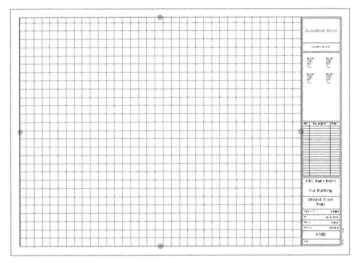

- To insert a view inside a sheet, locate it in the Project Browser, click it once, and drag and drop it inside the sheet.
- If the frame of the view is bigger than the available space, you may need to check the following:
  - The scale of the view should be changed.
  - The crop region of the view is large and you need to resize it.
- The title line will appear with each view dropped in a sheet:
  - It can be moved up and down.
  - You can lengthen or shorten it by the two ends.
  - By default, Revit will display the name of the view, but you can show a different name using the Properties palette:

- Once you start inserting sections, elevations, and callouts, all the heads in the plans will be populated with the right sheet number, as shown below:

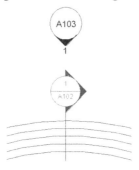

- *By default, you will deal with the view as a picture (you cannot edit its contents). If you need to work inside it, you need to activate it inside the sheet. Click the view inside the sheet and then right-click and select* **Activate View.** *When done, right-click and select* **Deactivate View**:

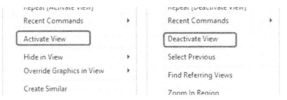

- *You can rotate a view inside a sheet. Click the view and set the rotation angle in the Options bar:*

- *The view cannot be inserted in several sheets.*

- *Once you add a view to any sheet and in the Project Browser you will see the white square at the left of the view name filled with a blue color.*

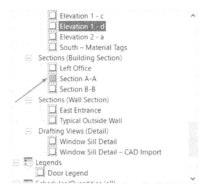

- Select any sheet, right-click it, you can Duplicate the sheet using the following options:
  - Duplicate Empty Sheet.
  - Duplicate with Sheet Detailing (this means if there are any detail elements in the original sheet, they will be copied).
  - Duplicate with Views: since you cannot insert the same view in multiple sheets, Revit will show the following message:

  - Revit will duplicate the view in order to be able to create a new sheet with a view, here, Revit is asking you what type of view duplication you want (discussed in Chapter 11).

- After you insert a view in a sheet, you can swap it with another view, do the following steps:
  - Select the view.
  - A context tab, will be shown like the following:

  - Select the new view name from the second pop-up list.

## PRINTING

- You can print sheets and views from your project.
- To print go to File menu and select **Print**, then **Print** option. You will see the following dialog box:

- Select the desired printer.

- Under **Print Range** select the **Selected views/sheets** option, then click the **Select** button and you will see the following:

- At the lower part of the dialog box under **Show**, click off **Views** to select only **Sheets**. The upper list will show only the available sheets in your model. At the right, click **Check All** to select all sheets as shown above, then click OK, and OK, to send these sheets to the printer.

## EXERCISE 13-5     CREATING SHEETS

**1.** Start Revit 2023.

**2.** Open the file **Exercise 13-5.rvt**.

**3.** Create the following sheets using A0 Metric title block (E 34 × 44 Horizontal.rfa) and insert inside them the related view (adjust the crop region when needed—you may need to shorten the section lines, and get the elevation bubble closer to the model):

   **a.** A100 – Ground Floor – Dimension (if there are trees, hide them)

   **b.** A101 – Section A-A

   **c.** A102 – Section B-B

   **d.** A103 – North Elevation

    **e.** A104 – Details (containing East Entrance, Left Office, 00 Ground-Toilets, and 00 Ground – Staircase & Elevator)

**4.** Check in the Project Browser the square next to the views used in this exercise, are they turned blue, or not?

**5.** Check the 00 Ground floor plan view; are bubbles of sections, elevations, and callouts filled?

**6.** Save the file and close it.

## NOTES

## CHAPTER REVIEW

1. In Walkthrough, you can change both Total Frames and Frames Per Second:

    **a.** True

    **b.** False

2. The Revit rendering engine is _____.

3. If you are inside a perspective view, you can switch between _____ and _____.

4. In printing, Title on Sheet is always equal to the name of the view, and you cannot change it:

    **a.** True

    **b.** False

5. In Solar Studies, one of the following is Not True:

    **a.** To get the right result, set the location and True North

    **b.** There are two types of Solar Studies

    **c.** Still, Single Day, and Multi Days are all types of Solar Studies

    **d.** You should show Sun Path and Shadows

6. To help reduce Render time, create Light Groups:

    **a.** True

    **b.** False

## CHAPTER REVIEW ANSWERS

1. a

3. Perspective, Isometric

5. b

## CHAPTER 14

# CREATING SCHEDULES

## This Chapter Contains

- Creating and modifying schedules

## CREATING SCHEDULES

- Schedules in Revit are data collected from the project file, whether they were counting schedules (door and window schedules) or quantity take-off for construction material.
- All the data is there, and all you need to do is put them in a nice, neat table. Since Revit is a total BIM solution, data that appear in the schedules are connected in two ways with the project data; any change in the project will reflect on the schedule, and vice versa.
- There are two ways to start the Schedule command:
  - Go to the Project Browser and locate a group called **Schedules/Quantities**; right-click this group to see the following list:

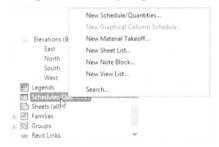

- Go to **View** tab, locate **Create** panel, and then select **Schedules** button to see the following list:

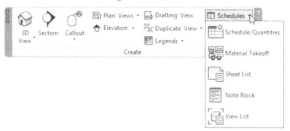

- In either way, select the first choice, which is **Schedule/Quantities**; you will see the following dialog box:

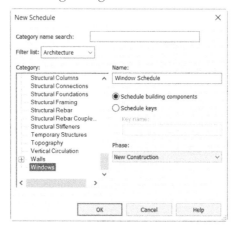

- The list of Categories is extensive, you can use a name to find the desired category you want.
- Using Filter list, uncheck everything except Architecture.
- At the left, select the desired Category like Doors, Windows, Furniture, etc. Type the name of your schedule, and then click OK.

- You will see the following dialog box:

- This dialog box contains all the necessary tools to create/control a schedule.
- This dialog box contains five tabs, they are: Fields, Filter, Sorting/Grouping, Formatting, and Appearance.
- In the coming pages, we will discuss each one of them.

## Fields

- In this tab, you will see the following:

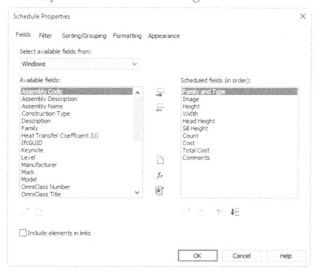

- In this tab, try the following:
  - Select the desired fields you want to include in your schedule from the available fields.

- Select one or more fields from the left (use [Ctrl] key) and click **Add parameter(s)** button.
- If you made a mistake and you want to remove one of the selected fields, select it and click **Remove parameter(s)** button.
- At the right, the list from top to bottom will be shown in the schedule from left to right. To rearrange the data, use **Move Up** and **Move Down** buttons.

### Filter

- In this tab, you will see the following:

- In this tab, you will set your filtering criteria:
  - Select the desired field (it should be one of the selected fields only):

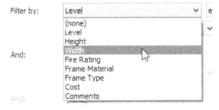

- Select the operator:

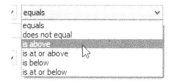

- Select the desired value (based on the selected field):

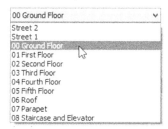

- If you want to make more than one filter condition, note that the only available option is "And."
- If you insert a schedule in a sheet, it will only show elements displayed in viewports on the sheet where the schedule is placed. For example, you want to insert the window schedule in the second floor plan, it will only show the windows related to second floor plan.

### Sorting/Grouping

- In this tab, you will see the following:

- You can sort/group by as many as four different selected fields. For each select to sort Ascending or Descending.
- Select whether to show the group name in the header and/or in the footer:

| A | B | C |
|---|---|---|
| Level | Family and Type | Height |

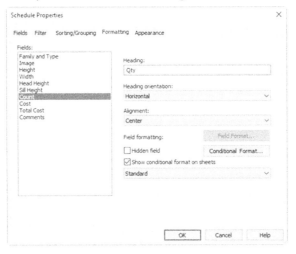

- Select whether to show total, count, and titles at the end of the table.
- Select whether to itemize every instance.

**Formatting**

- In this tab, you will see the following:

- For each field, you can do the following:
  - Type a new column heading other than the default.
  - Set heading orientation (Horizontal, Vertical).
  - Set alignment (Left, Center, Right).

- You can set the field format for the number fields and you will see the following:

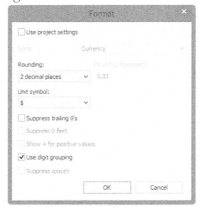

- You can select to hide the field.
- You can select to show a conditional format on sheets.
- You can select to show Standard number, Calculate totals, Calculate minimum, Calculate maximum, or Calculate minimum and maximum.

## Appearance

- In this tab, you will see the following:

- In this tab, try the following:
  - You can select to show/hide **Grid lines** and the **Outline** of the schedule along with specifying the line styles
  - Select whether to show the grid for headers/footers/spacers.
  - Select whether to show or hide the blank row before data.
  - Select whether to show or hide Stripe Rows and set the color.
  - Select whether to show or hide Stripe Rows on Sheets
  - Select whether to show or hide the Title and Headers.
  - Select the text type for title, headers, and body text.
- Once you click OK, a new schedule will be added to the Project Browser, as shown in the following:

## MODIFYING SCHEDULES

- When you visit the schedule view, you can resize columns by going to the line separating the two columns; click hold and drag the line to increase/decrease column width.
- Properties palette will be as follows:

- Under **Other**, you can see five buttons for the five tabs we discussed earlier. If you want to modify the schedule, simply click one of these buttons.
- While you are in the schedule view, there is much you can do:
    - Click on any row, and click **Highlight in Model** button to go directly to the view containing the element desired:

Highlight
in Model
Element

- To group several columns in a single title, select the column titles, then click **Group** button in **Titles and Headers** panel, and a new empty row above the column headers will appear; fill it with the proper group title, as shown in the following:

| Dimensions | | |
|---|---|---|
| Height | Width | Sill Height |
| 3550 mm | 1780 | 1 |

*While you are at the Schedule Views:*

- *The active row will be highlighted with light blue color*
- *You can freeze the Header of the schedule*
- *You can set the color of a column by selecting the header*

## INCLUDING AN IMAGE IN YOUR SCHEDULE

- To include an image for elements in a schedule, try the following:
    - Save the image(s) in a folder and remember the path.
    - Select one of the desired elements (door, window, etc.). At Properties, locate and click inside the **Image** field:

- Click the small button with the three dots and you will see the following:

- Click **Add** button and select the desired images.
- The following is the result:

- From the above picture, notice that Revit Architecture will remember the path of each image; changing the location of any image file would lead to losing it, hence, you will be asked to reload using the Reload button.
- You cannot see the image until you add the schedule view to a sheet.

## CALCULATED VALUE

- ▪ You can create a formula and add it as new field.
- ▪ To do that, do the following steps:
  - • In Schedule Properties dialog, go to **Fields** tab and click the **Add calculated Value** button.
  - • You will see the following dialog:

  - • Type a Name for the new field.
  - • Specify whether it is a Formula or Percentage.
  - • Specify Discipline (Common, Structural, HVAC, Electrical, Piping, Energy).
  - • Specify the field Type: whether it is Number, Text, and so on.
  - • Click the small button with the three dots near Formula to see the fields you can include in your formula:

  - • Select the desired fields, and add (*, /, +, -) between them.
  - • Click OK twice to end the command.

*You can use the keyboard to zoom in, zoom out, or retain the original zoom level of the schedule, using the following shortcuts:*
  ▪ *Use [Ctrl]+ (+) to zoom in to see a larger view of the text*
  ▪ *Use [Ctrl]+ (-) to zoom out to see a smaller view of the text*
  ▪ *Use [Ctrl] + (0) to reset to the original zoom level*

## EXERCISE 14-1   CREATING AND MODIFYING SCHEDULES

1. Start Revit 2023.

2. Open the file **Exercise 14-1.rvt**.

3. Go to 00 Ground floor plan.

4. Select one of the sliding windows. Under Identity Data, click inside the Image field, and click the small button with three dots. Using the **Add** button, add the three images in your exercise folder. Assign Sliding image to the selected window.

5. Start the Schedule and Quantities command. Using Filter list, uncheck everything except Architecture. From Category select Windows, and then click OK.

6. In the Fields tab, select the following fields: Family and Type, Image, Height, Width, Head Height, Sill Height, Count, Cost, and Comments. Click OK to end the command and see the result of the schedule.

7. To remove curtain wall windows from the schedule, use Filter and set the Sill Height > 0.0.

8. Freeze the schedule header

9. Set the cost for Archtop at 399.99, Sliding at 299.99, and Double Hung at 199.99. (When asked about "This change will be applied to all elements," click OK).

10. Use the zooming function to see the schedule text bigger, then smaller, then reset it to the original size

11. Under Sorting & Grouping, select Family and Type, click Header on, and click OK to see the results. Use Sorting & Grouping again, turn off Itemize every instance, and turn on Grand Total and set it to Totals only.

**12.** To get the Total Cost, do the following: go to Fields, click the Add calculated parameter button, and do as shown below:

**13.** Move Total Cost above Comments.

**14.** Go to Sorting and Grouping and click Footer for Family and Type.

**15.** Go to Formatting, select the Total Cost field, and change Standard to Calculate totals. Go to Fields Format, turn off Use project settings, choose Unit symbol to be $, and turn on Use digit grouping. Click OK twice.

**16.** Change the header of Count to be Qty, and Cost to be Unit Cost.

**17.** Go to Sorting & Grouping and turn off Header.

**18.** Select the four headings of *Height, Width, Head Height, and Sill Height*, and group them under the new header Dimensions.

**19.** Go to Formatting, select all columns except Family Type and Image, and set Alignment to Center.

**20.** Go to Appearance, turn on Grid in headers/footers/spacers, and turn off Blank row before data. Turn on Stripe Row in the view, and the sheet, and set the color to be dark gray

**21.** Set the Total Cost columns the Yellow shading

**22.** Assign an image for each window.

**23.** Using the Project Browser, right-click on Sheets (all), select the New Sheet option, and then click OK.

**24.** Drag and drop the schedule inside the sheet. The result should resemble the following for metric results:

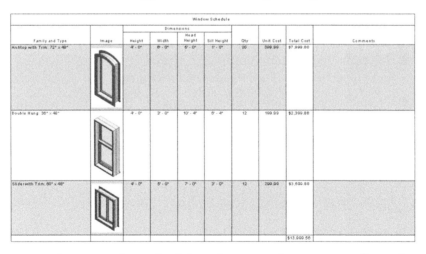

| Window Schedule | | | | | | | | | |
|---|---|---|---|---|---|---|---|---|---|
| | | Dimensions | | | | | | | |
| Family and Type | Image | Height | Width | Head Height | Sill Height | Qty | Unit Cost | Total Cost | Comments |
| M_Archtop with Trim: 1830 x 2438mm | | 2438 | 1830 | 2743 | 305 | 20 | 399.99 | $7,999.80 | |
| M_Double Hung: 0915 x 1220mm | | 1220 | 915 | 3120 | 1900 | 12 | 199.99 | $2,399.88 | |
| M_Sliding with Trim: 1830 x 1220mm | | 1219 | 1524 | 2134 | 914 | 12 | 299.99 | $3,599.88 | |
| | | | | | | | | $13,999.56 | |

**25.** The result should appear as the following for imperial results:

| Window Schedule | | | | | | | | | |
|---|---|---|---|---|---|---|---|---|---|
| | | Dimensions | | | | | | | |
| Family and Type | Image | Height | Width | Head Height | Sill Height | Qty | Unit Cost | Total Cost | Comments |
| Archtop with Trim: 72" x 48" | | 4' - 0" | 6' - 0" | 6' - 0" | 1' - 0" | 20 | 399.99 | $7,999.80 | |
| Double Hung: 36" x 48" | | 4' - 0" | 3' - 0" | 10' - 4" | 6' - 4" | 12 | 199.99 | $2,399.88 | |
| Slider with Trim: 60" x 48" | | 4' - 0" | 5' - 0" | 7' - 0" | 3' - 0" | 12 | 299.99 | $3,599.88 | |
| | | | | | | | | $13,999.56 | |

**26.** Duplicate Window Schedule and rename the new one All Windows.

**27.** Click Filter Edit button, and turn on Filter by Sheet, click Sorting/Grouping, and turn on Itemize every instance.

**28.** Create a new sheet based on A0 Metric (E34 × 44 Horizontal).

**29.** Insert in the new sheet, 00 Ground floor plan view at the top of the sheet.

**30.** Drag All Windows schedule and insert it (you may not see it at the beginning. If so, zoom out, and move it to the bottom of the sheet). You will notice only windows on the ground floor show up in the schedule.

**31.** Save and close the file.

## SCHEDULE KEY STYLES

- A key schedule will help you fill a finishing schedule quickly and easily.
- Create a key schedule using a key name, which will group finishes of a room (as an example) and then apply it to the normal schedule.
- To create a Schedule Key Style do the following:
  - Go to **View** tab, locate **Create** panel, expand **Schedules,** and click **Schedule/Quantities** option.
  - You will see the following dialog box; select a category, then select the Schedule keys option and type a Name for the key, as shown in the following:

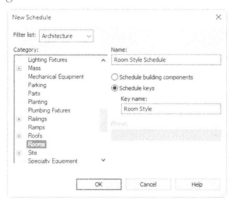

- Fields tab will include Key Name field automatically:

- Add the other desired fields. Then control Sorting/Grouping, Formatting, and Appearance.
- You will see an empty schedule. In the context tab, locate **Rows** panel, click **New Data Row** or right-click on the schedule and select **Insert Data Row** option.
- A new row will be added when default Key Name 1 is used. Add different values for the other fields:

| <Room Style Schedule> | | | |
|---|---|---|---|
| A | B | C | D |
| Key Name | Base Finish | Ceiling Finish | Wall Finish |
| 1 | Wood | Gypsum | Painting |
| 2 | Carpet | Gypsum | Wood |
| 3 | Tiles | Painting | Painting |

- When done, go to the schedule you would like to populate. Add a column to the schedule (such as the Room Style column). Once you input the key name, the columns will be filled automatically, as shown in the following:

| <Room Schedule> | | | | | | |
|---|---|---|---|---|---|---|
| A | B | C | D | E | F | G |
| Number | Name | Area | Room Style | Base Finish | Ceiling Finish | Wall Finish |
| 101 | Entry | 26 m² | 2 | Carpet | Gypsum | Wood |
| 102 | Stairs | 13 m² | 3 | Tiles | Painting | Painting |
| 103 | Entry | 36 m² | 2 | Carpet | Gypsum | Wood |
| 104 | Reception | 31 m² | 2 | Carpet | Gypsum | Wood |
| 105 | Room | 7 m² | 1 | Wood | Gypsum | Painting |
| 106 | Office | 45 m² | 1 | Wood | Gypsum | Painting |
| 107 | Patient | 11 m² | (none) | | | |
| 108 | Patient | 11 m² | (none) | | | |

## REUSING SCHEDULES

- You can reuse schedules you created in other projects. This is a very good tool to avoid repeating the same procedure in each project.
- To save a schedule to a file, try the following:
  - In Project browser, right-click on the schedule you want to save and select **Save to New File** option.
  - Using the Save As dialog box, specify a filename and a location. Click OK.
- To insert a schedule from another project, try the following:
  - Go to **Insert** tab, locate **Import** panel, expand **Insert from File,** and click **Insert Views from File** option:

  - In Open dialog box, select the project you want to use. Using the Insert Views dialog box, set the Views: to show schedules and reports only and select the schedule(s) you want to import into the current project, as shown in the following:

## EXERCISE 14-2   SCHEDULING KEY STYLES & REUSING SCHEDULES

**1.** Start Revit 2023.

**2.** Open the file **Exercise 14-2.rvt**.

**3.** Create a Schedule Keys for rooms, which appears as the following:

| A | B | C | D |
|---|---|---|---|
| Key Name | Floor Finish | Ceiling Finish | Wall Finish |

<Room Style Schedule>

| | | | |
|---|---|---|---|
| 1 | Tiles | Painting | Painting |
| 2 | Carpet | Gypsum | Painting |
| 3 | Wood | Gypsum | Wood |

**4.** Change the Room Schedule to look like the following (if you are using an imperial file, the only difference is the area value in m², or SF):

<Ground Floor Room Schedule>

| A | B | C | D | E | F | G |
|---|---|---|---|---|---|---|
| Number | Name | Area | Room Style | Floor Finish | Ceiling Finish | Wall Finish |
| 01 | Office | 20 m² | 3 | Wood | Gypsum | Wood |
| 02 | Office | 20 m² | 3 | Wood | Gypsum | Wood |
| 03 | Meeting Room | 11 m² | 3 | Wood | Gypsum | Wood |
| 04 | Reception | 11 m² | 2 | Carpet | Gypsum | Painting |
| 05 | Copy | 3 m² | 2 | Carpet | Gypsum | Painting |
| 06 | Storage | 4 m² | 1 | Tiles | Painting | Painting |
| 07 | Training | 7 m² | 3 | Wood | Gypsum | Wood |
| 08 | Toilet | 3 m² | 1 | Tiles | Painting | Painting |
| 09 | Toilet | 3 m² | 1 | Tiles | Painting | Painting |
| 10 | Kitchen | 3 m² | 1 | Tiles | Painting | Painting |
| 11 | Hall | 28 m² | 2 | Carpet | Gypsum | Painting |
| 12 | Office | 19 m² | 3 | Wood | Gypsum | Wood |
| 13 | Office | 19 m² | 3 | Wood | Gypsum | Wood |
| 14 | Meeting Room | 11 m² | 3 | Wood | Gypsum | Wood |
| 15 | Training | 11 m² | 3 | Wood | Gypsum | Wood |
| 16 | Storage | 3 m² | 1 | Tiles | Painting | Painting |
| 17 | Copy | 4 m² | 1 | Tiles | Painting | Painting |
| 18 | Reception | 8 m² | 2 | Carpet | Gypsum | Painting |
| 19 | Toilet | 4 m² | 1 | Tiles | Painting | Painting |
| 20 | Toilet | 4 m² | 1 | Tiles | Painting | Painting |
| 21 | Kitchen | 4 m² | 1 | Tiles | Painting | Painting |

**5.** Hide the Room Style column.

**6.** Save as the Room Schedule to a file to your exercise folder, naming it Room_Schedule_yourname.rvt.

**7.** Start a new project and import it to it. What exactly did you import (how many schedules?) and why? _____

**8.** Save and close the project.

## CREATING PROJECT PARAMETERS

- ▪ Project Parameter is the perfect way to add user-defined parameters.
- ▪ There are two types of pararmeters:
  - • Project parameters can be used in schedules, but not in tags.
  - • Shared parameters can be used in schedules and in tags. You can use them in other projects as well.
- ▪ To create a Project parameter, try the following:
  - • In Schedule Properties dialog box, in **Fields** tab, click **New Parameter**.
  - • You will see the following dialog box; select the **Project parameter** option, as shown below:

- • Under **Parameter Data** part, type in a Name for the parameter and select the Discipline, Group, and Type of Parameter in the drop-down lists. Choose whether it is an Instance or a Type:

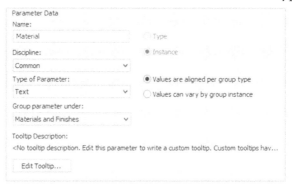

- • Instance parameters means for each instance of an element, this parameter will change. Type parameters means for all instances of an element, the parameter will be the same.
- • The Discipline can be set to: Common, Structural, or Electrical (in MEP also HVAC and Piping).

- Type of Parameter specifies how it will be stored inside Revit:

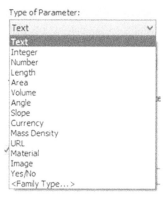

- The Group parameter under the drop-down will specify where this parameter will be shown in Properties and Family Types dialog boxes:

## CREATING FIELDS FROM FORMULAS

- Another way to create a new parameter is to create a formula. This is very popular in quantity take-off schedules.
- You can create either a formula or percentage.
- You have to add the right fields as your first step.

- To create a Calculated Value field, try the following:
  - In Schedule Properties dialog box, under **Fields** tab, click **Add calculated parameter** button. The Calculated Value dialog box opens as shown in the following:

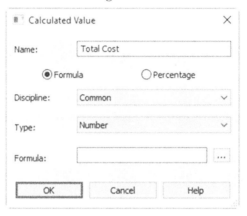

  - Type in a Name for the new field.
  - Specify Formula as the type of field.
  - Select the Discipline and Type as desired.
  - For the Formula, click the small button (with the three dots) to guarantee you get the exact field name, as shown below:

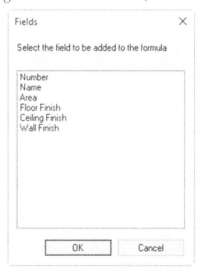

- Percentage fields can be used with area schedules.
- To create a Perecentage Field, try the following:

- Using Schedule Properties dialog box, under **Fields** tab, click **Add calculated parmeter** button. **Calculated Value** dialog box opens:

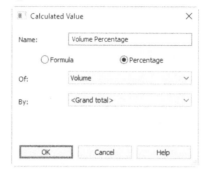

- Type a Name for the new field.
- Specify Percentage as the type of field.
- Select the field you want to take a percentage Of.
- Select the field you want to take a percentage By.
- The default value is <Grand total>, which means Revit will calculate the percentage based on the total of the entire schedule.

## CONDITIONAL FORMATTING

- Conditional formatting is to highlight a column in a different color in the schedule using condition(s).
- To create Conditional Formatting in a schedule, try the following:
  - Go to the desired schedule and click **Formatting** tab.
  - Select the desired field.
  - Click **Conditional Format** button.
  - You will see the following dialog box; select the Field, Test, and specify a Value, as shown in the following:

- Click on Background Color button and select a color to display in the schedule when the condition is satisfied. You can add more conditions.
- This is will be the result:

| A | B | C | D | E | F |
|---|---|---|---|---|---|
| Number | Name | Area | Floor Finish | Ceiling Finish | Wall Finish |
| | | | | | |
| 101 | Entry | 26 m² | Tiles | Painting | Painting |
| 102 | Stairs | 13 m² | Tiles | Painting | Painting |
| 103 | Entry | 36 m² | Tiles | Painting | Painting |
| 104 | Reception | 31 m² | Carpet | Gypsum | Painting |
| 105 | Room | 7 m² | Carpet | Gypsum | Painting |
| 106 | Office | 45 m² | Carpet | Gypsum | Painting |

**NOTE** *You can create conditional formatting based on Family, Type, and Family and Type.*

## CREATING MATERIAL TAKEOFF SCHEDULES

- You can create a special schedule, called Material Takeoff, which collects information about materials in a project and user-defined parameters.
- You can also create formulas or percentages based on material properties using the Calculated value option.
- To create a Material Takeoff Schedule, try the following:
  - Go to **View** tab, locate **Create** panel, expand **Schedules**, and click **Material Takeoff** option.
  - Using the dialog box, select **Category**. Type a Name and then click OK.
  - Using the Material Takeoff Properties dialog box, click the Fields tab and select the desired fields. You should include at least one of the Material fields (e.g., Material : Name, Material : Cost, etc.).
  - Use the Filter, Sorting / Grouping, Formatting, and Appearance tabs to control the schedule as we learned previously.

## SPLIT SCHEDULE ACROSS SHEETS

- If you have a schedule with lots of rows and one-sheet will not be able to accomedate it; you can split the schedule over muli sheets
- To do that, do the following steps:
  - Go to the desired schedule view
  - Locate Split panel and click Split & Place button

- You will see the following dialog box:

- Select the sheets you want to use
- Specify the Height on the Sheet, whether Split Evenly, or Custom
- Click Split & Place button

## CSV EXPORT FOR SCHEDULE

- You can export any schedule to a Dilemated file (*.CSV), which can be used by other software like Microsoft Excel
- You have to be in Schedule view, from **File** menu, select **Export**, select **Reports**, then select **Scheule**, when the dialog box appears, type in the name of the CSV file, once you click Save, you will receive the following dialog box:

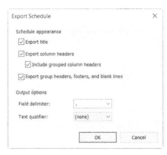

- Control the following:
  - Whether to export, Title, Column Headers including Grouped Column Headers, along with Group Headers, Footers, and Blank Lines.
  - Select the Field Delimiter comma, semicolon, tab, or space
  - Select desired Text qualifier whether none, ("), or (')
- Click OK for the export process to start

## EXERCISE 14-3   CREATING MATERIAL TAKEOFF SCHEDULES

1. Start Revit 2023.

2. Open the file **Exercise 14-3.rvt**.

3. Create a Material Takeoff Schedule.

4. Select Category = Floors, and leave the name without changing it.

5. Add the following fields: Level, Material:Name, Material:Area, Material:Volume, Material:Cost.

6. Go to Filter tab, and set Material:Name = Concrete, Cast In Situ (Concrete, Cast-in-Place gray).

7. Click OK to display the schedule.

8. You noticed that Material:Cost = zero.

9. Go to **Manage** tab, locate **Settings** panel, click **Materials** button, search for Concrete, Cast In Situ (Concrete, Cast-in-Place gray) material, go to the **Identity** tab, and set the price to 100 (10).

10. From Properties, click Fields tab, click Add calculated parameter button, and type in "Total Cost" as the name. Discipline = Common, Type = Currency. Type the following formula: (Material:Volume/1) * Material:Cost.

11. Click OK twice to display the schedule.

12. To set the right format for the Total Cost column, go to Properties and click the Formatting button. From the left select Total Cost, click the Field Format button, and click off the Use project settings checkbox. Set Rounding= 2 decimal places, Unit symbol = $, and Use digit grouping = on; click OK.

13. Click the Conditional Formatting button, set Total Cost < 20000 (Total Cost < 75000), and set the background color to be Cyan. Click OK.

14. Select Calculate Totals from the list.

15. Go to the Sorting/Grouping tab, switch on Grand Totals, and set it to Totals only. Click OK to display the schedule.

16. From Properties, click Fields, click the Add calculated parameters button, and name it "Volume Percentage"; it is Percentage of Material: Volume, By Grand total.

17. Move it up to be at the right of Material:Volume and format it to look like: 22.91%.

18. Start a new Material Takeoff schedule.

19. Category = Walls; call the schedule "CMU for External Walls."

20. Using the Fields tab, add the following fields: Function, Material:Name, Material:Area.

21. Use the Filter set Function = Exterior, and Material = Concrete Masonry Units.

22. Look at your schedule.

23. In order to calculate the number of CMU in one square meter, we have to multiply the area by 12.5, then round it off (there are 112.5 per 100 SF).

24. Using Properties, click the Fields button, click the Add calculated parameter button, and set the name to be CMU. It is Common, and it is Integer. The formula will be (Material:Area/1) *12.5 (for imperial use the following formula: (Material:Area/100) * 112.5)).

25. Go to Formatting tab, select CMU from left, and select Calculate totals.

26. Go to Sorting / Grouping tab, check off Itemize every instance, click on Grand Total, and set it to be Totals only.

27. How many CMU do you need in order to build the outside wall? _____ (38563 or 63154)

28. Go to Door Schedule, click Split & Place button, when the dialog box appears select A100, and A101, and select Split Evenly. Click Split and Place

29. Place the schedule to the right of the plan

**30.** Zoom in, you will notice that some of First Floor doors are shown in the Ground Floor sheet, click the schedule and show only the ground floor doors

**31.** Go to Window Schedule, click File / Export / Reports / Schedule, don't change the name of the CSV file

**32.** Do not change the default values, except, Field Delimiter = Space, and Text qualifier = (")

**33.** Save and close the file.

## NOTES

## CHAPTER REVIEW

**1.** In the Filter tab, you can use both And / Or between filters:

    **a.** True

    **b.** False

**2.** Calculated Value can be found in _____ tab.

**3.** While you are in the Schedule view:

    **a.** You can highlight any element in the model

    **b.** Group headers in one header

    **c.** You can modify the header of a column

    **c.** All of the above

**4.** You can group headers in schedule:

    **a.** True

    **b.** False

**5.** One of the following statements is NOT TRUE:

    **a.** In the Fields tab, you can include or exclude fields

    **b.** In the Sorting / Grouping tab, you can use Header and Footer

    **c.** In the Filter tab, you can specify many conditions

    **d.** In the Appearance tab, you can calculate the total of a column

**6.** In the Formatting tab, the button _____ can control whether to add a $ to the number or not.

## CHAPTER REVIEW ANSWERS

**1.** b

**3.** d

**5.** d

## CHAPTER 15

# PROJECT PHASING, DESIGN OPTIONS, AND PATH OF TRAVEL

## This Chapter Contains

- Project Phases
- Design Options
- Path of Travel

## PROJECT PHASING

- If your project is divided into phases, or you have renovations or additions to your existing building, then you need this feature in Revit.
- You can define existing, demolished, and new phases, and more.
- You can apply phases to elements and control them using views.
- There will always be the current phase, and when you draw a new element, it will be assigned to it automatically.
- Go to Properties, make sure there are no elements selected, scroll down to the Phasing area, and you will see the following:

| Depth Clipping | No clip |
|---|---|
| **Phasing** | ❯ |
| Phase Filter | Show All |
| Phase | New Construction |

- Set the **Phase** for elements created in the view (by default there are two phases, Existing or New Construction—the default phase).
- Set the Phase Filter to determine which phases are displayed in the view.
- For example, if Phase 2 is current, the **Show Complete** filter displays all construction up to Phase 2. **Show All** means it will display all phases up to the current phase with all except the current phase grayed out.

**NOTE**

- *Use the Duplicate view command to create duplicates of a view, and then apply different phase filters to each view.*
- *For example, you may want to show existing and demolished in one view and existing and new items in another view.*

### Phases and Elements

- To assign/modify phases for elements, you should select them first, check Properties, and you will see the following:

- All system and component elements will have phases.
- By default, Phase Created is set to New Construction and Phase Demolished is set to None unless you change them.
- On the other hand, you cannot assign phases for annotation elements (e.g., dimensions, text, and tags), view elements (e.g., sections, elevations, and callouts) and datum elements (e.g., grids and levels).
- You should use **Filter** in the context tab if you select multiple elements seeking to assign a phase for them, in order to rule out annotation, view, and datum elements.

### How to Demolish Elements

- To demolish an element, try the following:
  • Go to **Modify** tab, locate **Geometry** Panel, and select **Demolish** button.

- Select the elements to mark as demolished in the current phase.
- Press [Esc] to end the command.
  - If you demolish a wall, the doors or the windows associated with it will be demolished as well.
  - To reverse the process, select the demolished element, go to Properties, and change Phase Demolished to be None.
  - Curtain Walls, Panels, Grids, and Mullions cannot be assigned to a phase but the wall type can be.

### Creating Phases

- To create and manipulate phases, try the following:
- Go to **Manage** tab, locate **Phasing** panel, click **Phases** button:

- You will see the following dialog box; go to **Project Phases** tab and you will see the existing phases. There are two phases, Existing and New Construction, as shown in the following:

- Select the Phase Filters tab. Several phase filters are available, however, you can add more:

- For each filter select whether you want to assign By Category, Not Displayed, or Overridden, for New, Existing, Demolished, or Temporary.
- To understand the Overridden, select the Graphic Overrides tab. Set up the overrides as needed:

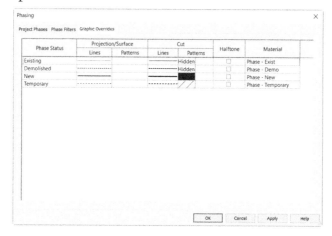

- As for Patterns, you can set the Foreground and Background as shown below:

- Normally you will set the Background as solid colored pattern and the foreground to hatched color
- To end the command, click OK.

## EXERCISE 15-1   USING PHASES

**1.** Start Revit 2023.

**2.** Open the file **Exercise 15-1.rvt**.

**3.** Make sure nothing is selected, go to Properties, and check that Phase Filter = Show All, and Phase = New Construction. This means the entire building was created in New Construction phase.

**4.** Select all elements (you can use Filter to filter out tags), and change the Phase Created to be Existing. The whole building will turn to a gray color.

**5.** Press [Esc] twice to make sure nothing is selected; at Properties change the Phase to be Existing.

**6.** Create a duplicate with detailing from Level 1 and call it Level 1 – Existing.

**7.** Go to the Phases command and add a new phase between Existing and New Construction and call it Demolished.

**8.** Create a duplicate with detailing from Level 1 – Existing and rename it Level 1 – Demolished.

**9.** Change the Phase to be Demolished.

**10.** Using the Demolish command, demolish all interior walls.

**11.** Select one of the interior walls, what is the Phase Created, and what is Phase Demolished.

**12.** Create a duplicate with detailing from Level 1 – Demolished and rename it Level 1 – New.

**13.** Change the Phase to be New Construction.

**14.** Add a new interior wall using Blockwork – 100 (Generic 8" Masonry) with radius = 6000 (20'-0"), add a door, and a table with chairs as shown in the following:

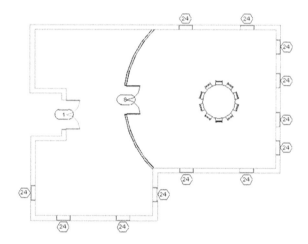

**15.** Go to Properties and change the Phase Filter to Show Complete.

**16.** Start Phase command, and select Phase Filters tab. The second row (Show Complete) change New to Overridden

**17.** Go to Graphics Override and change the Cut Pattern to look like below:

**18.** Zoom into the new wall and see how it changed

**19.** Save and close.

## DESIGN OPTIONS

- Design Options allow you to create many layouts for your model.
- Offer your client several interior wall designs, several roof designs, and so on.
- You need to create a Design Option Set and include underneath it the desired Design Options.
- To start the Design Options command, go to **Manage** tab, locate **Design Options** panel, and click **Design Options** button:

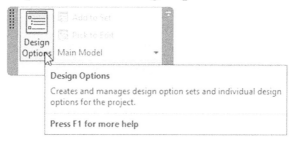

- Main Model—as shown in the following—is the part of your model that will not be affected by options. Create Option Sets and then add several Options per set as displayed:

### How to Set Up Design Options

- Under Option Set, click **New** button. An option set named Option Set 1 will be created along with a matching option as shown in the following:

- Use **Rename** button under Option Set to rename the default name.
- Click Options 1 (primary) and rename it as well using the **Rename** button under Option.
- You can create as many Option Sets/Options as you wish.
- When done, close the dialog box.
- Set the current Design Option Using **Manage** tab, or in Status Bar as shown in the following:

### How to Add Existing Elements to Design Options

- Normally you start adding elements after you define Design Options, but what if you have already elements in your model and you want to add them to the Design Option? The first thing is you must select elements before you activate a Design Option. Do the following:
- Make sure that Main Model is the Active Design Option.
- Select the elements that you want to add in a Design Option.

- Go to **Manage** tab, locate **Design Options** panel, and click **Add to Set** button:

- The **Add to Design Option Set** dialog box will be displayed. From the **Add selection to** drop-down list, select the Design Option Set. Then select the option(s) as shown in the following:

- When done, click OK. The elements are added to the option(s) and will not be modified in the main model.
- In typical views, only elements in the primary Design Option will be displayed, but they cannot be selected. In order to select these elements, you turn off the **Exclude Options** checkbox in the Status bar before selecting:

### How to Add New Elements to a Design Option

- In order to add new elements to a Design Option, try the following:
  - Set as current the desired Design Option as shown in the following:

- Main Model elements will be grayed out but new elements will be displayed in black.
- Add or modify elements as you wish.
- Finally, set the Active Design Option to Main Model.

### How to View Design Options

- The best way to present your Design Options is to create views and display each Design Option(s) in a separate view.
- To accomplish this, try the following:
  - Create a view (Floor Plan, Elevation, Section, or 3D view).
  - Start Visibility/Graphic dialog box.
  - Go to the Design Options tab and you will see the Design Option Set you previously created. Using the drop-down list for each Design Option Set, select the desired Design Option. The default value is <Automatic>, which means using the primary option:

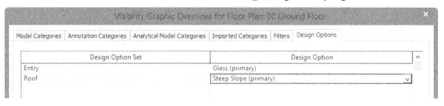

- Click **Pick to Edit** button in order to open a Design Option by selecting an element contained in the option. Go to **Manage** tab, locate **Design Options** panel, and click **Pick to Edit** button:

## Delete Design Options

- Once you present your design options to the client and agree on one of them, you can delete all other options in the project.
- To accomplish this, try the following:
  - Set Main Model to be the Active Design Option.
  - Open the Design Options dialog box.
  - Select the option you want to keep and click **Make Primary** button.
  - Select the Option Set and click **Accept Primary** button.
  - An alert box opens as shown below, warning you that all secondary options are going to be deleted. Click Yes if you are sure:

- If there are views associated with the deleted option, Revit will ask you to delete the associated views.
- Close the Design Options dialog box.

## EXERCISE 15-2   DESIGN OPTIONS

**1.** Start Revit 2023.

**2.** Open the file **Exercise 15-2.rvt**.

**3.** Go to Level 1 floor plan view.

**4.** Start the Design Options command.

**5.** In the Design Options dialog box, create the following Option Sets and Options, as shown in the following. When done, click the Close button:

**6.** Select Layout 1 from the Storage Area set. What happens to the rest of the model? _____ (grayed out)

**7.** Zoom into the lower left part of the building, start the Wall command, select the wall type Interior – Blockwork 100 (Generic 8"), set from Level 1 to level 2 minus 200 (0'-8"), draw the following walls (exact dimension of walls is not important), and then input the doors:

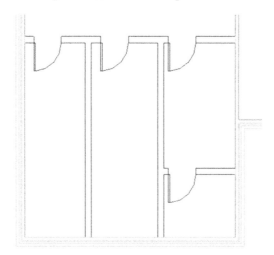

**8.** Select Layout 2 from the Storage Area set.

**9.** You will notice that whatever you did was removed.

**10.** Draw the following walls using the same settings, and add two doors as shown in the following (exact dimension of walls is not important):

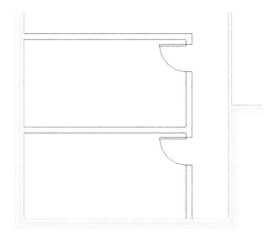

**11.** Now switch between Layout 1 and Layout 2.

**12.** Go to the Main Model.

**13.** Go to the Level 2 view.

**14.** Select Flat from the Roof set.

**15.** Start the Roof by Footprint command, turn off both Define Slope and Extend to wall core. Use Basic Roof – Generic – 125mm (Generic 9").

**16.** Using the [Tab] key select all outside walls (make sure the magenta line is at the inside edge of the wall).

**17.** Go to the Parapet floor plan view.

**18.** Select Gable from the Roof set.

**19.** Start the Roof by Footprint command, turn on Define Slope, and set Overhang = 500mm (0'-20"); select all the outside edges of the outside walls.

**20.** Go to the 3D view, and switch between Flat and Gable.

**21.** Create two duplicates of Level 1 and rename them Level 1 – Layout 1 and Level 1 – Layout 2.

22. Go to Level 1 – Layout 1, using Visibility/Graphics, go to Design Options tab, and set Storage Area = Layout 1. Do the same with Level 1 – Layout 2.

23. Create two duplicates of the 3D view, rename them 3D – Flat Roof and 3D – Gable Roof. Using Visibility/Graphics, set the two options for the two views.

24. Save and close the file.

## PATH OF TRAVEL

- Revit Path of Travel function will allow you to analyze travel distances and times between two points of your choice in the model.
- You can use this to determine the best escape plan if fire occurs in the building,
- In a plan view a path of travel is created by picking a start point and an end point. The model is analyzed and a path of travel is generated based on the model elements acting as obstacles (like furniture) along the path of travel.
- The path of travel calculated will avoid contact with model elements in the analysis zone and calculate the shortest distance between the start and end points.
- Path of travel elements can be seen in the current view only and act as detail elements.

**NOTE**
- *A path of travel element will not consider stairs and ramps and only calculates a horizontal distance on the plane of the path of travel. Hence, you cannot calculate path of travel between floors,*

- To start the command, go **Analyze** tab, locate **Route Analysis** panel, click **Path of Travel** button:

**Path of Travel**

Creates a path of travel along the shortest distance between 2 selected points.

**Press F1 for more help**

- Specify the first point, then specify the second point, Revit will generate the shortest path of travel, using the default line style "Path of Travel Lines." It will resemble the following:

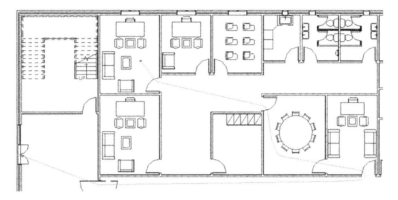

- Once the path of travel is generated, select it, and check the Properties, you will see the following:

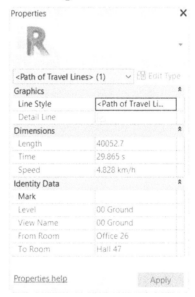

- You will find information, such as:
  - Length
  - Time
  - Speed
  - Level

- View Name
- From Room and To Room

■ If for any reason Revit could not create a Path of Travel, you will see the following message:

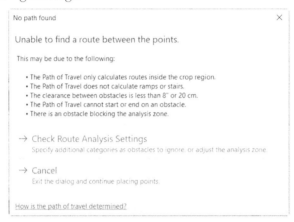

■ Revit will list the possible reasons why it could not create the Path of Travel. Along with the chance to alter the Route Analysis Settings or to cancel the process

■ By default, all elements are considered to be obstacles except doors. If you want to change this setting, go **Analyze** tab, locate **Route Analysis** panel, click **Route Analysis Settings** button:

- You will see the following dialog box:

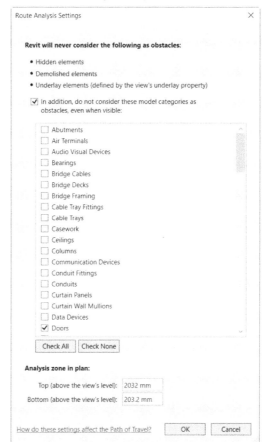

- From the dialog box, you can see that Revit will not consider hidden elements, demolished elements, or underlay elements to be obstacles
- You can see that all elements are considered obstacles except doors, you can turn on any other element to avoid considered it as obstacle
- At the bottom of the dialog box, Revit tells you that any obstacle below 203.2mm, or higher than 2032.0mm will not consider an obstacle

## Tagging

- Although we will discuss Tagging in Chapter 18, you still can tag the path of travel at the moment of creation

- Once you start the command, at the context tab, you can see the following:

- Make sure that the button labeled **Tag on Placement** is turned on
- You will see the following:

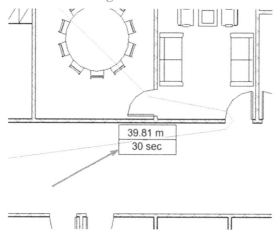

### Scheduling Path of Travel

- To do a schedule for the path of travel, simply select the path, using Properties, type in the name of the path in the Mark parameter, as shown in the following:

- Start Schedule / Quantities command, from Category select Lines, then Path of Travel option
- Select the desired parameters and create the schedule

## Reveal Obstacles

- Revit allows you to paint the obstacles by a different color so it will be easily identified.
- Go to **Analyze** tab, locate **Route Analysis** panel, click **Reveal Obstacles** button:

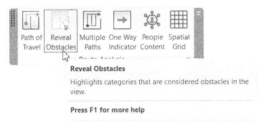

- Automatically you will see the obstacle elements displayed in orange color

## Add/Delete Waypoint

- Select the Path of Travel, at the context tab, you will see the following:

- You can add or delete Waypoints to the existing Path of Travel
- If you add or delete Waypoints, or you changed Route Analysis Settings, click **Update** button to see the new results after the changes occured

### Multiple Paths

- You can add multiple paths to study the different routes people may take
- Go to **Analyze** tab, locate **Route Analysis** panel, click **Multiple Paths** button:

- You will see the following dialog box:

- Input the minimum path separation between the multiple paths, when done click OK. Specify the first point and the second point, Revit will suggest – if possible, different paths calculated based on your data
- You will see the following:

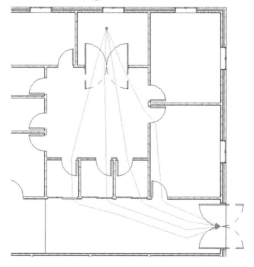

- Also, a message will appear as shown in the following:

### One Way Indicator

- You can assign a door to open in one way
- Go to **Analyze** tab, locate **Route Analysis** panel, click **One Way Indicator** button:

- Click the desired door, you will see the following:

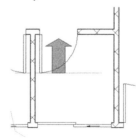

### People Content

- You can insert a person and physical distance radius into the view, to treat as an obstacle
- Go to **Analyze** tab, locate **Route Analysis** panel, click **People Content** button:

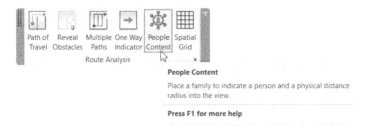

- Insert the person family, you will see the following:

**Spatial Grid**

- To place a square or hexagonal grid in a room element
- Go to **Analyze** tab, locate **Route Analysis** panel, click **Spatial Grid** button:

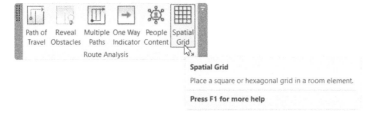

- You will see the following dialog box:

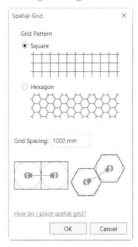

- Do the following:
  - Select which Grid Pattern; Square or Hexagon
  - Specify the Grid Distance
- Click OK when done
- Select a defined room to apply the selected spatial grid, from the Options bar, click Finish to end the command

## EXERCISE 15-3   PATH OF TRAVEL

**1.** Start Revit 2023.

**2.** Open the file **Exercise 15-3.rvt**.

**3.** Start Path of Travel command, make sure that Tag on Placement is turned on

**4.** For the first point click inside the upper left room (near the stair-case), for the second point, click the outside exit door at the top. Press [Esc] twice to end the command

**5.** From the tag what is distance and what is the time? _____, _____ (40.50m, 30 sec) (132'-7", 30 sec)

**6.** Run the command again, specify the same first point, for the second point, specify the exit at the right. Press [Esc] to end the command

**7.** From the tag what is the distance and what is the time? _____, _____ (31.97m, 23 sec) (104'-7", 23 sec)

**8.** Select the first path and second path, what is the speed for both paths? _____ (4.828 km/h) (3 mph)

**9.** Turn on Reveal Obstacles

**10.** Start Route Analysis Settings, and turn on Furniture, click OK

**11.** Using Update key, update the two paths. How this affect the distance and the time for both paths? _____ (they are less)

**12.** Using Properties, and Mark parameter, name the first path Left Exit, and the second path Right Exit

**13.** Create a schedule to compare the first path distance and time to the second path, using the following Fields: Mark, Length, Time

**14.** Hide the two paths (use Hide Elements and not Hide Category).

**15.** Start Multiple Paths command and set the minimum path separation to 1000mm (3'), select the first point to the upper right empty room with double door, and the second point to the right exit. How many paths generated? _____ (2)

**16.** Delete the two paths

**17.** Redo the command with minimum separation distance = 500mm (1'6")

**18.** How many paths generated? _____ (4)

**19.** Delete all the generated paths

**20.** Start One Way Indicator command, and assign it to the door nearest to the right exit

**21.** Redo the Multiple Paths command, for the same two points. How many paths generated? _____ (3)

**22.** Delete all the generated paths

**23.** Start command Spatial Grid, choose Square and 1000mm (3'), and insert it into the space between the offices

**24.** Start People Content command, using the spatial grid, insert two persons at the middle of the room beside each other

**25.** Redo the Multiple Paths command, for the same two points. How many paths generated? _____ (2)

**26.** Save and close

# NOTES

## CHAPTER REVIEW

1. There are two premade phases in all Revit files:
   a. True
   b. False

2. You can control which Design Option is displayed in the current view using _____ dialog box.

3. You can find the demolish command in:
   a. Modify tab, Geometry panel
   b. Architecture tab, Geometry panel
   c. Manage tab, Phasing panel
   d. Modify tab, Phasing panel

4. One Design option will be primary in one Option Set:
   a. True
   b. False

5. One of the following statements is NOT TRUE:
   a. For each element you can assign a phase created in
   b. For each element you can assign a phase demolished in
   c. You can assign a phase for annotation elements
   d. Both a & b are true

6. While creating Design options, _____ is the part of your model which will not be affected by options.

7. Show All and Show Complete are _____.

8. By default, for Path of Travel, all elements are considered obstacles except _____.

## CHAPTER REVIEW ANSWERS

1. a
3. a
5. c
7. Phase Filter

CHAPTER 16

# TOPOSURFACES IN REVIT

**This Chapter Contains**

- Creating Toposurfaces
- Project Base and Survey Point
- Editing Toposurfaces
- Site Settings
- Property Lines and Building Pads
- Modifying Toposurfaces
- Annotating Site Plans
- Site Components

## INTRODUCTION

- In this chapter, we will discuss topography or toposurface as named in Revit.
- You can create and modify:
  - Project Base Point and Survey Point, which will affect the accurate coordinates, elevation, and True North angle of the project
  - Toposurfaces
  - Site settings that affect the toposurfaces

## PROJECT BASE POINT AND SURVEY POINT

- Every project has a Project Base Point and a Survey Point, as shown in the following:

- You can see both points when you go to the **Site** plan view.

### Project Base Point

- Revit does not deal with a coordinate system like AutoCAD, which has WCS and UCS.
- Project Base Point describes the origin (0,0,0) of the project coordinate system.
- It influences the absolute elevation of your model.
- To modify the Project Base Point, click on the icon (circle with X inside it) and modify N/S, E/W, Elev, and Angle to True North.
- Another way is to input the values in Properties as shown in the following:

- Make sure that your model is less than 20 miles (32 Km) from the location of the project base point if you want to accurately deal with Revit.

### Survey Point

- Survey Point is a specific point in the real world, such as a survey marker.

- It is used to orient the building when shared with a program that uses a different coordinate system (like Civil 3D software, for instance).
- Unclip Survey Point first to specify its coordinates.

*To display your model with True North, try the following:*
- *Duplicate the plan view that you want to rotate to true north (most likely the Site plan view).*
- *Go to Properties of the true north plan and set the Orientation to True North, as shown in the following:*

**NOTE**

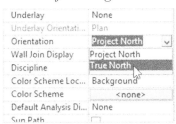

- *Accordingly, the view rotates to show True North specified in the Project Base Point.*

## CREATING TOPOSURFACES

- You can create a toposurface using one of the following methods:
  - Inputting points manually
  - Importing a CAD file with 3D points
  - Importing a points file which contains X, Y, Z points

### Inputting Points Manually

- To create a toposurface by inputting points, try the following:
  - Open a Site plan view or 3D view.
  - Go to **Massing & Site** tab, locate **Model Site** panel, and click **Topo-surface** button:

- In the context tab, locate **Tools** panel and click **Place Point**.
- In Options Bar, set Elevation for a point.
- For the first three points, you can select Absolute Elevation. Then you can also select Relative to Surface:

- Click the drawing area to place points.
- Continue placing points. You can change the Elevation of points as needed.
- You will see contours forming after three points with the same height.

**How to Create a Toposurface Using an Imported File**

- To create a toposurface using an imported file, try the following:
  - In a Site plan view or 3D view, *import or link* the file (DWG, DXF, or DGN) that holds the 3D points and objects like polylines or splines.
  - Go to **Massing & Site** tab, locate **Model Site** panel, and click **Topo-surface**.
  - In the context tab locate the **Tools** panel, expand **Create from Import** and click **Select Import Instance**.
  - Select the imported file by hovering over one of the lines, as shown:

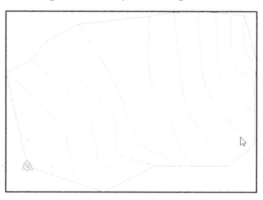

- You will see the following dialog box; select the layer(s) that hold the points and the objects, as shown in the following:

- Click OK. The new toposurface is created with points at the same elevations as the imported information.
- Click (✔) to end the command.

■ You may see the following warning; simply ignore it:

■ AutoCAD points and objects become Revit toposurfaces, as you will see:

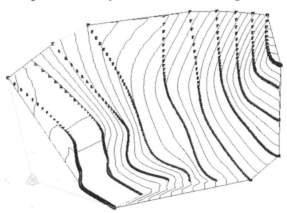

**How to Create a Toposurface from a Points File**

■ To create a toposurface from a Points File, do the following steps:
  • Open the Site plan view or 3D view.
  • Go to **Massing & Site** tab, locate **Model Site** panel, and click **Toposurface**.
  • In the context tab locate **Tools** panel, expand **Create from Import**, and click **Specify Points File** button.
  • You will see the Open dialog box; select the CSV or comma delimited text file which contains the list of points (N/S, E/W, Elev) and click Open.
  • The Format dialog box will appear; pick the units of your file and click OK. Pick one of the following: Decimal feet, Decimal inches, Meters, Centimeters, or Millimeters:

  • The points will create a toposurface. Click (✓) to end the command.

## EDITING TOPOSURFACES

■ You can make changes to a toposurface by adding points or editing existing point locations and elevations. You can also modify the Properties of a toposurface, including material and phasing information.
■ To edit a toposurface, try the following:
  • Select the toposurface that you want to edit.
  • In the context tab locate **Surface** panel and click **Edit Surface** button:

- **Edit Surface** context tab will appear; locate **Tools** panel and click **Place Point** to add more points to your surface.
- To edit existing points, select one or more points, either to edit their elevation value, as shown in the following, or to delete the point using [Del] on the keyboard.

Boundary Point | Elevation: 3048.0

- Click (✓) to end the command.

- *To reduce points and to speed up the working process, use **Simplify Surface** command.*

- *Still in **Edit Surface** context tab, locate **Tools** panel and click **Simplify Surface** to reduce the number of points. Set the required accuracy, as shown in the following:*

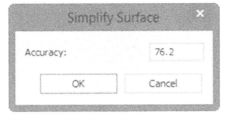

## SITE SETTINGS

- To change the site settings, go to **Massing & Site** tab, locate **Model Site** panel, and click **Site Settings** as shown:

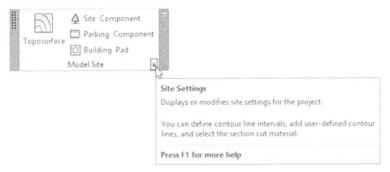

- You will see Site Settings dialog box, which allows you to control how contours are displayed in both plan and section views, as shown in the following:

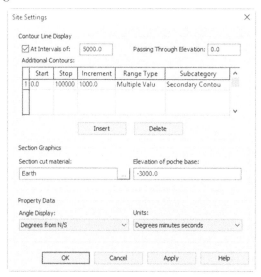

- Under Contour Line Display:
  - **At Intervals of**: this value will specify the distance between the primary contour line and another. They will display in a heavy line style.
  - **Passing Through Elevation**: For primary contour lines, this value will specify the starting elevation for them. Example: if you type 5, instead of the default value 0, then the primary contour lines will pass through 5, 5005, 10005, and so on.
- Under Additional Contours:
  - **Start / Stop**: Specify the start and stop of additional contour lines.
  - **Increment**: If Range Type is Multiple Value, then Increment means the distance between an additional contour and another.
  - **Range Type**: Single Value means you can specify the location of the single contour in the Start Value; Increments are not available. Normally we want to display a group of additional contours.
  - **Subcategory**: to specify the object style of the additional contours. Pick from Hidden, Primary, Secondary (thin lines), or Triangulation Edges.
  - **Insert or Delete buttons**: to insert or delete additional contour definitions.

- Under Section Graphics:
  - **Section Cut material**: Select the three small dots button to pick the desired material which will be displayed when you create a section in the toposurface. Normally, you will choose one of the following materials: Earth, Sand, Grass, Asphalt, or Water.
  - **Elevation of poche base**: Specify the height of the poche (poche means the hatch representing the earth) in case of section, usually negative value.
- Under Property Data:
  - **Angle Display**: For Property Line (discussed shortly) select how angles will be displayed: Degrees, or Degrees from N/S.
  - **Units**: For Property Line, select angle units to be displayed: Decimal Degrees or Degrees Minutes Seconds.

## CREATING PROPERTY LINES

- After placing your toposurface, you should add both the property lines and building pad, as shown below:

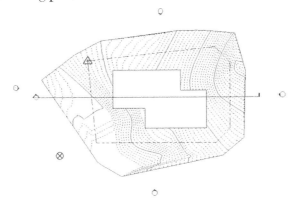

- To create Property Lines, try the following:
  - Go to **Massing & Site** tab, locate **Modify Site** panel, and click **Property Line** button:

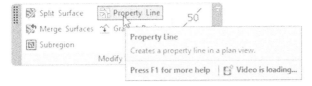

- You will see the following dialog box; select how you want to create the property line:

- If you select **Create by sketching**, use the available sketch tools to create the property line, and click (✓) to finish.
- If you select **Create by entering distances and bearings**, input Distance and Bearing relating to N/S, and E/W, then specify the type of object, whether Line or Arc, as shown below; then click OK to finish:

## CREATING BUILDING PADS

- A building pad is a sketched shape representing the building in the topo-surface, hence, cutting it and removing earth out of the designated area. The following image shows a section, in a building pad cutting the topo-surface.

- To create a Building Pad, try the following:
  - Open one of the Site plan views. You should have toposurface available to start sketching the building pad.
  - Go to **Massing & Site** tab, locate **Model Site** panel, and click **Building Pad** button:

  - Using Draw panel in the context tab, select Boundary Line then use Pick Walls or any sketch tools to draw the building pad. Use reference planes and temporary dimensions to help you sketch it precisely.
  - Using Properties, specify a Level and a Height Offset from Level for the building pad and set the phasing as needed. To finish Click (✓).

- A Building pad boundary, like any other boundary in Revit, must be continuous, closed, with no overlaps. If you have multiple buildings, create a pad for each one.
- Use the Slope Arrow to slope pads in one direction for drainage.

**NOTE** *Use the Label Contours command to display the elevation of the contour lines after creating them.*

- Go to **Massing & Site** tab, locate **Modify Site** panel, and click **Label Contour** button:

- Contour labels "keep readable" from any direction.

## EXERCISE 16-1    CREATING TOPOGRAPHICAL SURFACES

**1.** Start Revit 2023.

**2.** Open the file **Exercise 16-1.rvt**.

**3.** Open Site floor plan view.

**4.** Use the Import CAD command and select SitePlan.dwg in your class folder. This file was created using AutoCAD. Set Colors to Black and White, and Positioning to Auto - Origin to Origin (for imperial set Import units = feet) and click Open.

**5.** Double-click the mouse wheel to zoom all. As you can see the two origins coincide.

**6.** Click on the Project Base Point and change the information in Properties as listed in the following:

    **a.** N/S: 17733 (59'-0")

    **b.** E/W: 40536 (135'-0")

    **c.** Elev: 703010 (2343'-0")

    **d.** Angle to True = 15 degrees

7. Double-click the mouse wheel to see all the imported toposurface.

8. Select the Survey Point and unclip it, and input the following data in Properties:

    **a.** N/S: 96197 (1587'-0")

    **b.** E/W: 278081 (4836'-0")

    **c.** Elev: 620000 (2066'-0")

9. Reclip the Survey Point.

10. Change the scale of Site plan view to 1:500 (1" = 40'-0").

11. Using Duplicate with Detailing, create a duplicate view of the Site plan view and rename it Site-True North. Use Properties to set the Orientation to True North.

12. Open the Site plan view.

13. In **Massing & Site** tab locate **Model Site** panel and click **Toposurface** button. In the context tab locate **Tools** panel, expand **Create from Import**, and click **Select Import Instance**.

14. Select the imported CAD file.

15. In the Add Points from Selected Layers dialog box, click the **Check None** button. Select the layer **Existing Site** and click OK.

16. In the Surface panel, click (✓).

17. Hide the imported CAD file and the elevation markers.

18. Start the **Site Settings** dialog box.

19. Set At Intervals = 5000 (10'-0"), set the Increment = 500 (10'-0"), and Subcategory= Hidden Lines, then click OK.

20. Zoom in to see the primary contour lines and the secondary contour lines.

21. Using Label Contours try to label all possible contour lines (for imperial, duplicate the existing family, and change the Text size = ½").

22. Create a horizontal and vertical Section through the site.

23. Rename the section as Horizontal Site Section and Vertical Site Section.

24. Open the Horizontal Site Section view.

**25.** Hide the CAD Import.

**26.** The material displayed in the section is Earth, which was specified in the Site Settings. But you can see that Earth is not covering the whole topo-surface.

**27.** Go to Site Settings again, change the value for Elevation of poche base to be -8000. Now the Earth hatch is correctly set.

**28.** Draw a Property Line similar (not necessarily exactly) to the following:

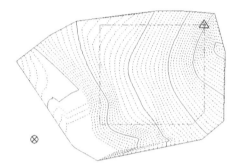

**29.** Create Building Pad similar to the following:

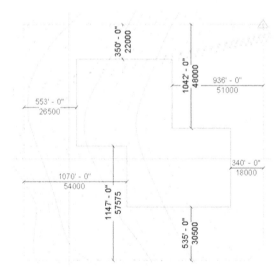

**30.** In Properties, verify that the Level is Level 1 and the Height Offset from Level is equal to 0. Click (✔) to finish the command.

**31.** Visit the two sections to see how the building pad is cutting the toposur-face.

**32.** Visit the 3D view. Hide the CAD Import. See how the building pad is set.

**33.** Save and close the project.

## EDITING TOPOSURFACES

- Normally you will have the existing toposurface. But you will need Revit to help you create the proposed toposurface, which will be your final product.
- Using commands like Subregion, Split Surface, and Graded Region will allow you to design drainage away from your building and specify the car parking lots and other utilities.

### Creating Subregions

- Subregions will help you specify materials for parts of the toposurface without changing the existing contour lines.
- To create a subregion, try the following:
  - Go to **Massing & Site** tab, locate **Modify Site** panel, and click **Subregion** button:

  - In context tab, use Draw tools to sketch the borders of a subregion.
  - Using Properties, set the Material for the subregion as desired.
  - Click (✓) to finish the command.
  - To modify an existing subregion, click its border; in the context tab, locate the Subregion panel then click the Edit Boundary button. You will be back to sketch mode to make the necessary changes.
  - To delete an existing subregion, select it and press the [Del] key.

### Splitting Surfaces

- Split Surface command will really split toposurface into discrete parts. You can modify and assign material for each one by itself.

- To split a surface, try the following:
  - Go to **Massing & Site** tab, locate **Modify Site** panel, and click **Split Surface** button:

- Select the desired toposurface.
- In the context tab use the Draw tools to draw the borders of the splitting boundary. Each split process should end up with two separate boundaries. It is accepted in this type of boundary to be open.
- Using Properties, set the Material for the surface as desired.
- Click (✓) to finish the command.

## Merging Surfaces

- If you split surfaces and want to undo your work, you can use Merge Surfaces to finish the job. The only condition is they should be adjacent to each other.
- To join surfaces, try the following:
  - Go to **Massing & Site** tab, locate **Modify Site** panel, and click **Merge Surfaces** button:

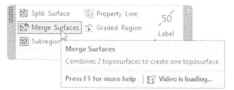

- Select the two adjacent surfaces in no particular order.
- Material of the first picked surface will be applied to the newly created surface.

## Grading a Site

- Revit does not have the best grading tool comparing to other specialized software, but this command is good enough for architects.
- Use this command to change the nature of the existing surface by changing the elevation of the points to match your intended design.

- Do not forget to change the phase of the existing surface before you proceed with this command. Set the Phase Created of the surface to be Existing.
- Revit automatically calculates the Cut and Fill for the surface. This information is displayed in Properties.
- To grade a site, try the following:
  - Make sure that Phase Created of the surface is Existing.
  - Go to **Massing & Site** tab, locate **Modify Site** panel, and click **Graded Region** button:

  - You will see the following dialog box. It will inform you that the existing toposurface is demolished (that is why we set the Phase Created to Existing) and a matching toposurface was created in the current phase and asks how you want the new toposurface to be copied. The options are either to create a copy of the original, or a copy based on the perimeter points only:

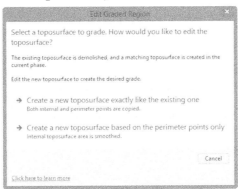

  - Select the desired toposurface.
  - In the context tab use the tools to add more points, delete points, or set new elevation for existing points.
  - Click (✓) to finish the command.

## ANNOTATING TOPOSURFACES

- In order to label a toposurface, use commands like: Label Contours (already discussed), Spot Elevation, and Spot Coordinate; add to that normal commands like Dimension and Text.

### Spot Elevations and Spot Coordinates

- Spot Elevation displays the elevation of a selected point, while Spot Coordinate displays the coordinate data of the project:

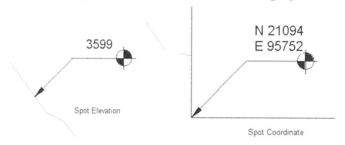

- To add Spot Elevations / Coordinates, try the following:
  - Go to **Annotate** tab, locate **Dimension** panel, and click Spot Elevation or Spot Coordinate buttons.

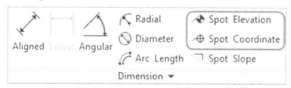

  - Using Properties, select the type that you want to use.
  - Using Options Bar, select whether to show Leader and Shoulder. For Spot Elevation, specify the Display Elevations option as shown:

  - Select a point to dimension and place the leader line and target or text. Revit automatically displays the information from the selected point. Click Modify or press [Esc] to end the command.

## SITE AND PARKING COMPONENTS

- As your final touches, you can add trees and other landscaping shapes, along with parking spaces. These are just like other components we added previously but these will be placed on the contours and not on the levels.
- To access site and parking components, go to the **Massing & Site** tab, locate **Model Site** panel, and then click **Site Component** button or **Parking Component** button:

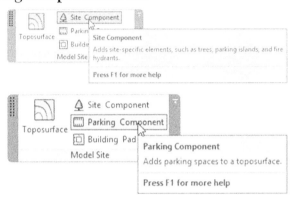

- To add parking-related RFA files, go to the Site / Parking folder; you will find direction arrows, parking islands, curbs, parking spaces, and so on.
- In the Site / Accessories folder you can find bollards and benches. In the Site/Logistics folder you can find trucks, cranes, and so on. In the Site / Utility folder you can find cable connection, catch basins, manholes, and so forth.
- To add more trees to your project, go to the **Planting** folder. You will see plants, trees, and shrubs. Each one of these encompasses an assortment of types with different heights.

## EXERCISE 16-2    MODIFYING TOPOSURFACES

**1.** Start Revit 2023.

**2.** Open the file **Exercise 16-2.rvt**.

**3.** Go to the Site floor plan view.

**4.** Hide the two sections.

**5.** Select the toposurface and change Phase Created to Existing.

**6.** Using the **Split Surface** command, create the two surfaces as shown in the following (create each one in a separate command):

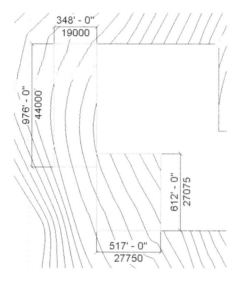

**7.** Using Visual Style, change the view to Shaded.

**8.** Start the Graded Region command, and in the Edit Graded Region dialog box, select Create a new toposurface exactly like the existing one, and select the top left surface (which is intended to be car park).

**9.** Press [Esc] twice to get out of adding new points mode. Delete all points, except those on the four corners of the surface.

**10.** Select all of the corner points. In the Options Bar, set the Elevation to 3500 (72'-0"). If you have the view shaded, the cut areas display in red.

**11.** Click (✔) to finish the command.

**12.** Start the Graded Region again and select Create a new toposurface based on the perimeter points only. Select the lower surface and use the procedure that you used with the car park: delete all of the points except the corner points and set those points to 6500 (136'-0").

**13.** Click (✔) to finish the command.

**14.** Go to the 3D view, and you can see both the existing and the proposed.

**15.** Select one of the proposed surfaces and change the Phase Created to Existing; you will notice that in Properties, under Others, Revit will show you the Net cut/fill, Fill, and Cut volumes.

**16.** Check the other surface.

**17.** Change back the Phase Created to be New Construction for both surfaces.

**18.** Go to the Site plan view.

**19.** Change the Visual Style to be Hidden.

**20.** Change the Phase Filter to be Show Complete.

**21.** Using Spot Coordinate (display elevation) command add two spot coordinates, one at the Project Base Point and one at the Survey Point.

**22.** Using Spot Elevation, show the spot elevation for the lower left corner of both newly created surfaces.

**23.** Using the Parking Component, add 4800 × 2400mm – 60 deg (9' × 18' – 60 deg) to fill in the right side of the parking space (if the added parking component was not displayed, then click Pick New Host from the context tab, and click on the newly created surface).

**24.** Add trees and other site components to the Site plan view.

**25.** Save and close the file.

# NOTES

## CHAPTER REVIEW

1. Split Surface and Subregion commands have the same effect:

   **a.** True

   **b.** False

2. After the Graded Region command, Revit will calculate _____ volumes.

3. Concerning toposurface creation, which is true?

   **a.** You can import a CAD file to create toposurface out of it

   **b.** You can use a text file containing 3D points to create a toposurface from it

   **c.** You can manually input points after setting elevation for each point

   **d.** All of the above

4. You cannot label toposurface:

   **a.** True

   **b.** False

5. One of the following statements is NOT TRUE:

   **a.** You can create a Property Line using two methods

   **b.** You can create a Building Pad

   **c.** You can add trees and site components to my project

   **d.** You cannot merge regions after they were split

6. _____ describes the origin (0,0,0) of the project coordinate system.

## CHAPTER REVIEW ANSWERS

**1.** b

**3.** d

**5.** d

## CHAPTER 17

# CREATING ROOMS AND AREAS

## This Chapter Contains

- Creating Rooms
- Creating Areas
- Creating Color Schemes

## INTRODUCTION TO CREATING ROOMS

- If you have bounding walls (curtain walls as well) you can create a room element. Revit will give it a name and a number, and you can modify both.
- Use the Room Separator Line to create rooms in open areas.
- Revit will calculate Area and Perimeter by default. To calculate Volume, Revit needs Floor, Roof, or Ceiling.
- Later you can create schedules for Rooms, accordingly room finishing, and quantity take-off.
- Also, you can create Color Scheme plans.

## CREATE AND MODIFYING ROOMS

### To Create Rooms One by One

- ▪ To create rooms one by one, do the following steps:
  - • Go to **Architecture** tab, locate **Room & Area** panel, and click **Room** button:

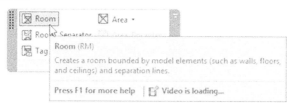

  - • Using the context tab click **Tag on Placement** to add a tag when placing a room.
  - • Using the Options Bar, set the Upper Limit. The default Offset value is the height of the level. This will control the volume calculations, as shown previously:

  - • Also, you can set how the tag will be placed; you have three options, Horizontal, Vertical, and Model (this tag can be rotated to be aligned with walls and other boundaries). Set whether you want a Leader with the tag or not.
  - • Set whether the room to be placed will be a New room requiring a new name, or a repetition of an existing room; select the appropriate choice from the drop-down list (you should create a schedule containing all your rooms with their associated numbers).
  - • Click inside the desired bounded area to place the room:

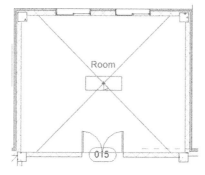

  - • Continue adding rooms.

### To Create All Rooms in One Step

- ■ To create all rooms in one step, try the following:
  - • Go to **Architecture** tab, locate **Room & Area** panel, and click **Room** button:

  - • Using the context tab click **Place Rooms Automatically** and all rooms will be added; you will see the following message:

  - • Click **Close**. You will see all rooms were added with default name and default number using the settings preset in the Option bar.

### Room Separation Lines

- ■ You can use Room Separation lines to define rooms. If you have an open area for offices without having real wall to separate them, you can use this tool to create rooms without walls:

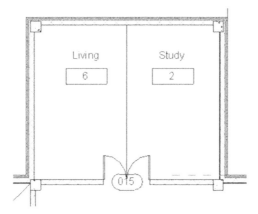

- To create a room separation line, do the following steps:
  - Go to **Architecture** tab, locate **Room & Area** panel, and click **Room Separator** button:

  - Using the Draw panel in context tab, draw the desired separator.
  - Use the Room command to add the desired room in the areas bounded by the separation lines.

### Modifying Rooms

- To modify a room, select it by hovering over the boundary until you see the crossing mark, pick it and modify its Properties.
- Room Properties include Constraints (the Upper Limit and Limit Offset and Base Offset), Dimension information (read only data — like Area, Perimeter, Volume, etc.), and Identity Data (like Number, Name, Occupancy):

- To edit information for multiple rooms, select them and then change the information using Properties.

### Adding Room Tags

- Add Room Tags while placing rooms or afterword. There are three room tag types: Room Tag, Room Tag With Area, and Room Tag With Volume:

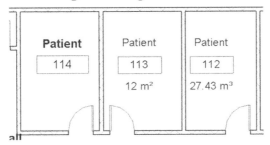

- To place a room tag, do the following steps:
  - Place rooms without tags in your project.
  - Go to **Architecture** tab, locate **Room & Area** panel, expand **Room Tag** and click **Room Tag** button:

  - From Properties, select the tag type. Click inside the room you want to tag. Repeat the process with other rooms, and press [Esc] to end the command. Rooms can be tagged in *plan* or *section* views.
  - Use Properties to change the room name or number, or select the tag and click he name or number to edit them. Or you can set a new tag orientation or show/hide the leader.
  - You can delete the tag at any moment, and tag it again later on.
  - Do not move any tag outside its room without a leader.

## ROOM SCHEDULE

- Create a room schedule showing information (same as showing in Properties) saved in rooms. You can edit them in both places as needed:

| Room Schedule | | | |
|---|---|---|---|
| Number | Name | Area | Occupancy |
| 101 | Entry | 26 m² | |
| 102 | Stairs | 13 m² | |
| 103 | Entry | 36 m² | |
| 104 | Reception | 31 m² | |
| 105 | Room | 7 m² | |
| 106 | Office | 45 m² | |

- To create a new Room Schedule, do the following steps:
  - Start the Schedule command (discussed previously). From **Category**, select **Rooms**, type the name of the new schedule, and click **OK**.
  - Using **Fields** tab, add the desired fields.
  - Using **Filter** tab, set criteria to filter your data.
  - Using **Sorting/Grouping** tab, set the criteria to sort and group data.
  - Using **Formatting** tab, format each column in your schedule.
  - Using **Appearance** tab, set the lines and fonts.

### Room Settings

- You can set the way Revit calculates volumes and boundary locations.
- Go to **Architecture** tab, locate **Room & Area** panel, expand panel title, and click **Area and Volume Computations** button:

- You will see the following dialog box:

- Under Room Area Computation, select one of the following: At wall finish (default), At wall center, At wall core layer, or At wall core center.

## EXERCISE 17-1   CREATING ROOMS

**1.** Start Revit 2023.

**2.** Open the file **Exercise 17-1.rvt**.

**3.** Make sure you are at the 00 Ground floor plan.

**4.** This floor plan contains four locations; two small, and two big offices.

**5.** Working at the lower left location, start Room command.

**6.** Using the image below, place the first room and press [Esc].

**7.** Set room number to be 01 and do not change the name of the room; add the eleven rooms as shown below and then change their names:

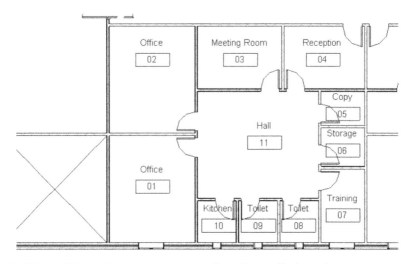

**8.** Using Place Rooms Automatically, place all the other rooms and name them as shown below (you may receive a different result concerning the room numbers, which is acceptable):

**9.** Using the Room Separator, create a new room in the Hall (in the upper image Room # 23) and call it Display Area, and renumber it to be 23a, as shown below (or use the number you have in your project):

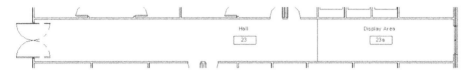

**10.** Unhide the section at the top of the plan. Go to the Section 1 view.

**11.** Go to **Architecture** tab, locate the **Room & Area** panel, expand **Tag Room** button, select the **Tag All Not Tagged** command, select **Room Tags**, and click OK. All rooms in 00 Ground are tagged. Select all tags, and change the family to M_Room Tag Room Tag with Area (Room Tag Room Tag with Area).

**12.** Start the Schedule command and create a new Room Schedule as shown (Call it Ground Floor Room Schedule) using the following fields:

| <Ground Floor Room Schedule> | | |
|---|---|---|
| A | B | C |
| Number | Name | Area |
| 01 | Office | 20 m² |
| 02 | Office | 20 m² |
| 03 | Meeting Room | 11 m² |
| 04 | Reception | 11 m² |
| 05 | Copy | 3 m² |

**13.** Save and close the file.

## CREATING AREAS

- You can define two types of areas in Revit: Gross Building and Rentable.
- Gross Area is the total area of the building, and it is measured from the outer face of the external walls. You can have for each level a separate gross area plan.
- Rentable Area is area measurements based on the standard method for measuring floor area in office buildings.
- There are three steps you will do:
  - Create Area Plan
  - Create Area Boundaries
  - Create Areas

### Creating Area Plan

■ To create an Area Plan, do the following steps:

- Go to **Architecture** tab, locate **Room & Area** panel, expand **Area**, and click **Area Plan** button:

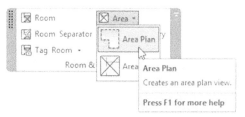

- You will see the following dialog box; select the type of area plan, either Gross Building or Rentable:

- Select one or more level(s) for which you want to create new views. Each level has an area plan created for it.
- Revit will show the following message, asking to automatically create area boundary lines associated with all external walls as shown below:

- Click Yes. In your Project Browser, you will see that area plans are created for the desired levels, as shown in the following illustration. Revit will create Gross Building views separate from Rentable views:

- For Gross Building, boundary lines are created on the outside edge of the exterior walls, whereas for Rentable area plans, they are created on the inside edge of the exterior walls.

### Placing Area Boundaries

- The next step is to create area boundary lines in the Rentable area plan to define parts of the building, such as offices, stores, and common areas.
- To create Area Boundary lines, do the following steps:
  - Go to the Rentable Area Plan you want to work in.
  - Go to **Architecture** tab, locate **Room & Area** panel, and click **Area Boundary Line** button:

- Using the context tab, select the desired sketch tool:
  - Using Pick Lines, Revit will show in the Option bar a checkbox called **Apply Area Rules**, which means you will add boundary lines from the faces or center of the selected wall according to the type of area you are specifying.
  - Using Line or Rectangle tools, boundary lines will stay as drawn regardless of area types you apply to them.

### Adding Areas

- This is the third and final step of Area definition process.
- Adding areas is similar to adding rooms.

- Using Properties, you can specify the area type, which controls the location of the boundary lines.
- For instance, if you are working in the Rentable Area Plan, the Office area type will select the center of the wall, the major vertical penetration uses the outside edge of the wall, whereas the building common area used the inside edge of the wall.
- To add an area, do the following steps:
  - Go to **Architecture** tab, locate **Room & Area** panel, expand **Area** button, and click **Area** button:

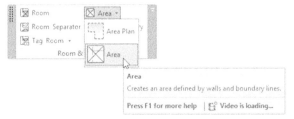

  - Before you place the area, go to Properties and set the Area Type. Hover your mouse pointer over the closed area boundaries and click to add. The default tag will display the name of the area and the area value.
  - Press [Esc] to end the command.
- As we did with Room definition, you can select multiple areas to edit the name and number collectively.
- If in the future you changed the area boundaries or moved the walls connected to them, the calculated area would be changed accordingly.

### Tagging Areas

- Go to **Architecture** tab, locate **Room & Area** panel, expand **Tag Area** button, and click **Tag Area** button:

**NOTE** *You can create Gross Area or Rentable Area Schedules that include the area, name, and type, as we did with Room Schedule.*

## EXERCISE 17-2    CREATING AREAS

**1.** Start Revit 2023.

**2.** Open the file **Exercise 17-2.rvt**.

**3.** Start the Area Plan command, select type = Gross Building, select 00 Ground. Click Yes in the alert box to create boundary lines at the external walls. The gross area = 842 m² (9058 SF).

**4.** Using the Area Plan command create a **Rentable** area plan for 00 Ground. Click **Yes** at the prompt to create boundary lines at external walls. The area boundary is on the inside wall of the building this time.

**5.** Add area boundaries and areas (specify area type from Properties):

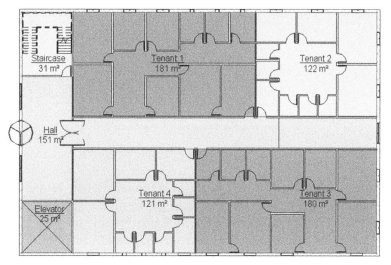

**6.** Using the Schedule command, create a Rentable Area Schedule. Show the name, area type, and area:

| <Area Schedule (Rentable)> | | |
| --- | --- | --- |
| A | B | C |
| Name | Area Type | Area |
| Tenant 1 | Office Area | 181 m² |
| Tenant 2 | Office Area | 122 m² |
| Tenant 3 | Office Area | 180 m² |
| Tenant 4 | Office Area | 121 m² |
| Hall | Building Common Area | 151 m² |
| Staircase | Major Vertical Penetration | 31 m² |
| Elevator | Major Vertical Penetration | 25 m² |

**7.** Save and close the file.

## CREATING COLOR SCHEMES

- ■ Create Color Schemes that show the various uses for rooms and areas.
- ■ You can show different color schemes according to Room/Area names, area value, and so on.
- ■ To create a color scheme in a view, do the following steps:
  - • Make sure you are in the right view (normally you need to duplicate the view and apply the color scheme to the duplicate).
  - • Using Properties, under **Graphics** locate the **Color Scheme** parameter, and click the button next to it, as shown:

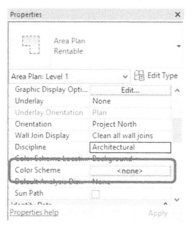

- • You will see the Edit Color Scheme dialog box; locate the Schemes area, and from Category select the desired category (normally if you are in Area plan rentable, your available choice is Areas Rentable). Select an existing scheme as shown below. Click OK. The new color scheme displays in the view.

- Add a legend that matches the Color Scheme. Go to **Annotate** tab, locate **Color Fill** panel, and click **Color Fill Legend** button:

- Add the legend in the desired place as shown:

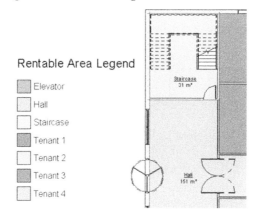

### Creating a New Color Scheme

- To create a new color scheme, do the following steps:
    - In the desired view, make sure that nothing is selected. Using Properties, locate **Color Scheme** and click the button next to it. Another way is to go to **Architecture** tab, locate **Room & Area** panel, expand the panel title, and click **Color Schemes** button:

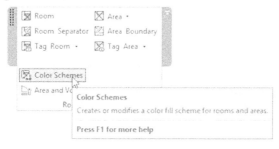

- Either way you will see the Edit Color Scheme dialog box; under Schemes, select a Category: Areas (Gross Building), Areas (Rentable), or Rooms, as shown:

- Select an existing scheme and click the Duplicate button:

- In the New color scheme dialog box, type the name of the new scheme.
- Type a Title for the new color scheme. This displays when the legend is placed in the view.
- Use the Color drop-down list to choose an existing color scheme, as shown:

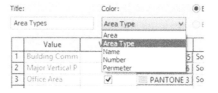

- Select whether you want the scheme to be By value or By range (discussed later).

- Use the green plus button to add more rows to the scheme, as shown. Control the visibility (Visible column), Color, and Fill Pattern as you wish. In Use is the read-only parameter:

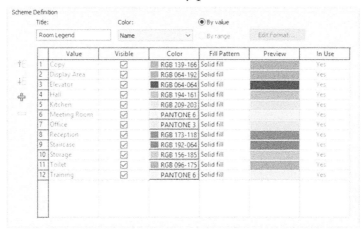

### Color Scheme By Value

- If you select By value option, value is assigned by the parameter data for the room or area object like Name, Number, Area, Parameter (these are for Rooms).
- Values are automatically updated when new data is introduced.
- By default, Revit will sort the data that will be displayed; to change this order, select the desired row and use arrow up and arrow down as shown:

- To remove a row, select it and click the red minus sign. This is only permissible if the data is not used in the room or area elements in the project.

### Color Scheme By Range

■ By range option means you have to specify a range (for example area values). Change the **At Least** variable and the Caption, as well as the visibility, color, and fill pattern, as shown in the following:

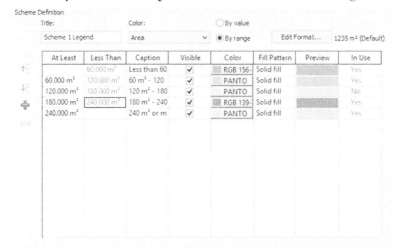

■ Only Area and Perimeter parameters can be set by range.
■ To change how the units are displayed, click Edit Format to control the units display format; you will see the following dialog box:

■ Click off the **Use project settings** checkbox and change the settings as you wish.

- To add a new row, make sure to select the row above the new row then click Add Value. The new row increments according to the previous distances set or double the value of the first row.

## EXERCISE 17-3    CREATING COLOR SCHEMES

**1.** Start Revit 2023.

**2.** Open the file **Exercise 17-3.rvt**.

**3.** Create a duplicate with detailing from 00 Ground and call it **00 Ground – Room Names Color Scheme**.

**4.** Make sure nothing is selected; from Properties, select Color Scheme.

**5.** Use the existing Name color scheme and click OK.

**6.** Add a legend to the right of the plan.

**7.** Create two duplicates with detailing only for the Area Rentable plan. Name the first one: **00 Ground Rentable Area Name** and the second one **00 Ground Rentable Area By Range**.

**8.** Go to 00 Ground Rentable Area Name, and start the color scheme command.

**9.** Select the existing scheme, and under Color, change the value to Name.

**10.** Click OK, and then add a legend.

**11.** Go to 00 Ground Rentable Area By Range, and start the color scheme command.

**12.** Do as in the following image (for imperial use 650 SF, 1300 SF, and 1900 SF):

Scheme Definition

| | Title: | | Color: | | ○ By value | | | |
| | Rentable Area Legend | | Area | ∨ | ◉ By range | Edit Format... | 1235 m² (Default) | |

| | At Least | Less Than | Caption | Visible | Color | Fill Pattern | Preview | In Use |
|---|---|---|---|---|---|---|---|---|
| ↑ | | 60.000 m² | Less than 60 m² | ☑ | RGB 156- | Solid fill | | Yes |
| | 60.000 m² | 120.000 m² | 60 m² - 120 m² | ☑ | PANTO | Solid fill | | No |
| ↓ | 120.000 m² | 180.000 m² | 120 m² - 180 m | ☑ | PANTO | Solid fill | | Yes |
| ➕ | 180.000 m² | | 180 m² or more | ☑ | RGB 139- | Solid fill | | Yes |

**13.** Go back to **00 Ground – Room Names Color Scheme**.

**14.** Change the name of one of the offices to Manager Room. You will notice that a new color was introduced, and the legend updates by adding a new value.

**15.** Save and close the file.

# NOTES

## CHAPTER REVIEW

**1.** You can define a room with or without walls:

    **a.** True

    **b.** False

**2.** The default room tag will show _____ and _____.

**3.** Area in Revit:

    **a.** Two types: Gross and Interchangeable

    **b.** Two types: Gross and Rentable

    **c.** Area is one type

    **d.** None of the above

**4.** You can create a Color Scheme By Range:

    **a.** True

    **b.** False

**5.** In a Color Scheme, if you create a room legend and later on add a new room, Revit will ask you if you want to add the new room to the legend:

    **a.** True

    **b.** True, but you need to use the Refresh button first

    **c.** False, you cannot add a room after the legend is created

    **d.** False

**6.** Without using walls, you can use _____ lines to define rooms.

## CHAPTER REVIEW ANSWERS

    **1.** a

    **3.** b

    **5.** d

# 18

# *TAGGING AND DETAILING*

## This Chapter Contains

- Adding tags
- Creating details

## INTRODUCTION

- In this chapter, you will learn how to:
  - Add tags for different Revit elements whether in 2D or 3D views
  - Create details using different methods, including importing ready-made AutoCAD details drawings

## HOW TO ADD TAGS IN REVIT

- Some elements such as Doors and Windows have a button while adding them, allowing you to place the tag with the element.
- Walls, on the other hand, and other Revit elements do not have this capability; hence, Revit provides you with three tools to add tags for any desired element, except Rooms which have separate commands as discussed previously.

- There are three ways to tag in Revit, they are:
  - Tag by Category
  - Multi-Category
  - Material

**Tag by Category**

- This method will tag individual elements according to their category. It will insert Window tags on window, and Furniture tags on furniture.
- The tag should be loaded prior to the command or Revit will produce the following message (user **Insert** tab, **Load from Library** panel, click **Load Autodesk Family** to load the tag you need)

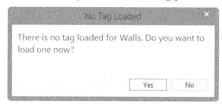

- Allowing you to load the desired tag family.
- To issue this command, go to **Annotate** tab, locate **Tag** panel, then click **Tag by Category** button:

- In Options bar, you will see the following:

- It contains the following:
  - Horizontal or Vertical: decides whether the text appearing as a tag will be displayed horizontally or vertically:

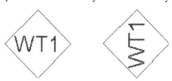

- Select whether you want the tag with leader or without a leader.
- Select whether you want the tag as attached end or free end:

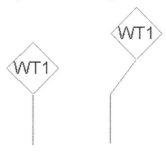

- For attached end, you can specify the length of the leader.
- After you specify these settings, simply click the desired element and a tag will be placed.
- To tag all the incidences of a certain element, you can use the Tag All command; this will add tags to all desired elements in the current view.
- To issue this command, go to **Annotate** tab, locate **Tag** panel, and then click **Tag All** button:

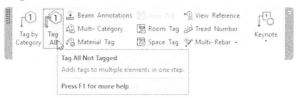

- You will see the following dialog box:

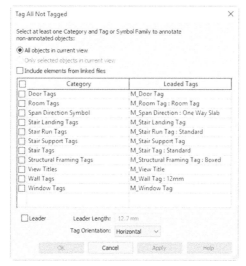

- This dialog box title is "Tag All Not Tagged." The first sentence at the top reads: "Select at least one Category and Tag Family to tag non-tagged objects." This is a self-explanatory sentence.
- The default option is to add tags for "All objects in current view." You can add to "Include elements from linked files."
- Select the desired tag(s) by clicking and holding [Ctrl] to select multiple tags. If you want all tags in this dialog box, click the checkbox near **Category**
- Select whether you want:
  - Leader, and if yes, what is the length of the leader
  - What is the tag text orientation: Horizontal or Vertical?
- Click OK to apply the command.

### Material Tag

- To add a material tag of the selected elements.
- To issue this command, go to **Annotate** tab, locate **Tag** panel, and then click **Material Tag** button:

- The same Option bar will be shown. Select the desired part of an element and click to specify the location of the tag. You will see the following:

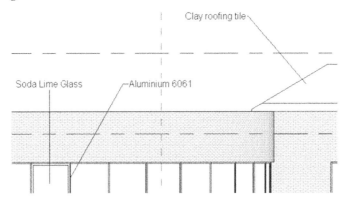

*If you want to tag in a 3D view, make sure of the following:*
- *Create a duplicate of the 3D view and give it a name.*
- *Orbit your model to see the desired side of your model.*
- *Using the View Control bar at the bottom, click the Unlocked 3D View button (we need to lock the 3D view as a first step, before adding tags in 3D):*

- *You will see the following choices:*

**NOTE**

- *Select the Save Orientation and Lock View option.*
- *Now you are free to tag on this 3D view, using any of the methods you learned earlier.*
- *For walls, you can input a tag, by doing the following steps:*
  - *Select one segment of the desired wall type, then click the Edit Type button.*
  - *You will see the Type Properties dialog box; browse until you see Type Mark, and then input the number you want:*

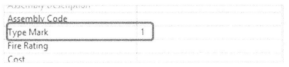

*If you are tagging similar items, instead of tag every single one, you can use a Multi-Leader Tag instead.*
- *Do the following steps:*
  - *Add a tag for an element*

**NOTE**
  - *Select that tag, check the context tab, and click Add / Remove Host button, as shown below:*

- *Select the new host and result will resemble the following:*

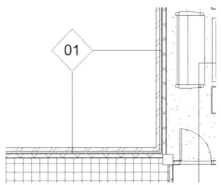

- *Continue adding more items for the same element*
- *You can control to Show All Leaders, Hide All Leaders, Show One Leader, Select Leaders to Show, and Merge Leaders*
- *Select the tag and check the Properties, you will see the following:*

| Graphics | |
|---|---|
| Leader Line | ☑ |
| Orientation | Model |
| Angle | 0.00° |
| Leader Type | Attached End |
| Host Count | 4 |

- *In the above example, you can see that there are 4 tags attached to the same host. Use the dialog box to turn on/off the leader, change the tag Orientation and angle, and finally change the leader Type*

## EXERCISE 18-1    ADDING TAGS

1. Start Revit 2023.

2. Open the file **Exercise 18-1.rvt**.

3. Duplicate the 00 Ground floor plan and name the duplicate 00 Ground – Wall Tagged.

4. Using the Tag By Category command, tag one of the outside walls at the left, bearing in mind the following:

   a. Horizontal

   b. Leader = On

    **c.** Leader = Attached End

    **d.** Distance = 20mm (3/4")

**5.** The tag will show a question mark; click the question mark and type 01. A message will appear saying: "You are changing a type parameter. This could affect many elements. Continue?" Click Yes.

**6.** Select any other outside wall, click the Edit Type button, and search for Type Mark, you will find it 01.

**7.** Click the tag you just added, from context tab, click Add / Remove Host, and select another adjacent wall

**8.** Repeat the same with other adjacent walls

**9.** Using the Tag All command, tag all walls using M_Wall Tag: 12mm (Wall Tag: ½") with leader.

**10.** Input Type Mark = 02 for walls in the inside, 03 for the elevator & staircase, 04 for curtain walls, 05 for Storefront, and 06 for Toilet walls.

**11.** Clean all redundant tags and try to create a neat floor plan with tags. If you need to rotate any tag, select it, from Properties change the Rotation angle

**12.** Duplicate the 3D view and call the new view 3D – Window Tagged.

**13.** Orbit your model to show some of the windows in the north side of the building.

**14.** Save Orientation and Lock View.

**15.** Tag All windows without leader; you will see the following:

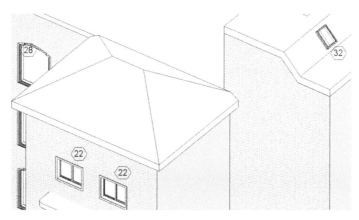

**16.** Go to the South elevation view.

**17.** Create a callout as shown in the following. Rename it South – Material Tags. Input the Material tags as shown:

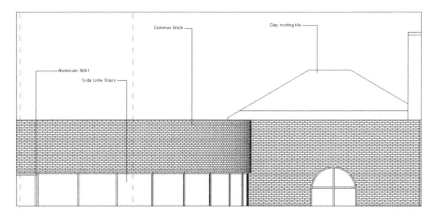

**18.** Go to South elevation view

**19.** Tag the some of the vertical Mullions

**20.** Save and close the file.

## CREATING DETAILS

- There is more than one way to create a detail in Revit. They are:
  - Create a detail view from scratch.
  - Use one of the callouts as your starting point, and then add to it the necessary adjustment to make it a detail.
  - Get an AutoCAD detail and convert it to Revit elements.
  - Get an AutoCAD detail and link it to the section view.

### Creating a Detail View from Scratch

- You need to create an empty view first. To do that, go to **View** tab, locate **Create** tab, and then click **Drafting View** button:

- You will see the following dialog box:

- Type the name of the new view, specify scale, and click OK. This will create a new empty view.
- Go to the **Annotate** tab, locate the **Detail** panel, and you will see the following:

- Use any of the following tools to draw your detail:
  - Detail Line, to draw lines, arcs, and so on using Line Style. This appears in the following context tab:

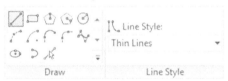

  - Revision Cloud
  - Region: there are two choices: Filled Region and Masking Region. Using Properties, select the desired pattern to be used.
  - For Component, you will see the following:

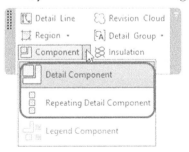

- Both the Detail and Repeating Detail Component will be selected from Properties. Some of the Detail Component is pre-loaded, but lots of them will need you to Load Family first.
- Go to the Detail Items folder and you will see the following:

| Name | Date modified | Type |
|------|---------------|------|
| Div 01-General | 6/18/2013 11:24 AM | File folder |
| Div 02-Sitework | 6/18/2013 11:24 AM | File folder |
| Div 03-Concrete | 6/18/2013 11:24 AM | File folder |
| Div 04-Masonry | 6/18/2013 11:24 AM | File folder |
| Div 05-Metals | 6/18/2013 11:24 AM | File folder |
| Div 06-Wood and Plastic | 6/18/2013 11:24 AM | File folder |
| Div 07-Thermal and Moisture Protection | 6/18/2013 11:24 AM | File folder |
| Div 08-Doors and Windows | 6/18/2013 11:24 AM | File folder |
| Div 09-Finishes | 6/18/2013 11:24 AM | File folder |
| Div 10-Specialties | 6/18/2013 11:24 AM | File folder |
| Div 11-Equipment | 6/18/2013 11:24 AM | File folder |
| Div 12-Furnishings | 6/18/2013 11:24 AM | File folder |
| Div 13-Special Construction | 6/18/2013 11:24 AM | File folder |
| Div 14-Conveying | 6/18/2013 11:24 AM | File folder |
| Div 15-Mechanical | 6/18/2013 11:24 AM | File folder |
| Div 16-Electrical | 6/18/2013 11:24 AM | File folder |

- That represents the 16-division system of Construction Specification Items (CSI). It contains an abundance of ready-made parts that will help you build your detail easily.
- Finally, Insulation, which will create an insulation pattern. This is good for sections.

### Callout or Section as Starting Point

- Another way to draw a detail is to create a callout or wall section, and then add to it any of the above-mentioned tools.
- Since the callout and the wall section will show part of the model, then you can use Keynotes.
- There are three types of Keynotes:
  - **Element Keynote**: which will tag the desired element with the keynote specified for the element
  - **Material Keynote**: which will tag the desired element with the keynote specified for the material existing in the element
  - **User Keynote**: which will tag the desired element with a tag you will select.

- To see a wall keynote (for instance), select one wall and click Edit Type. Under **Identity Data**, check the **Keynote** field. Click the small button with three dots inside it and you will see the following:

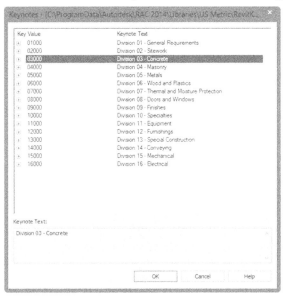

- Keep selecting the sub-categories to the point where you reach to what you want.
- The final value will be the following:

- If you go to the Material Browser, locate your desired material; at the right, click Identity tab. Under **Revit Annotation Information**, locate Keynote.
- Repeat the same procedure we did earlier. It will appear as the following:

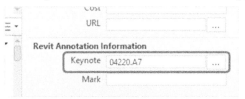

- To add a keynote to your model, go to the **Annotate** tab, locate the **Tag** panel, then click the **Keynote** button, and you will see the following:

- Below you will see two examples for the result after the Keynote is inserted:

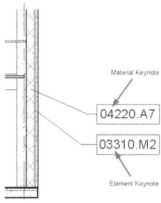

### Starting with an AutoCAD File

- Since there are millions of people who have used AutoCAD and are still using it, most likely all of the companies will have a huge library of details.
- Create a drafting view just we did above, then either link or import the AutoCAD file (we recommend importing).
- You have two paths to take:
  - Leave the file intact without any change and link it to one of the sections. To do the linking, start the Section command (Wall Section) at the context tab and you will see a **Reference other view** checkbox; click it, and then select the name of the drafting view containing your detail:

- This is the shape of the section:

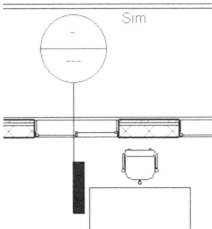

- The second path will be to import an AutoCAD drawing and then Explode it.
- After explosion, it will be converted to Revit elements. You will start to change text lines to line styles of your own. You can add any other detailing components like we learned about previously.

**NOTE** *In order to use the same detail in future projects, you have to save the view. To do that go to Application Menu, select* **Save As***, then* **Library***, and select* **View***:*

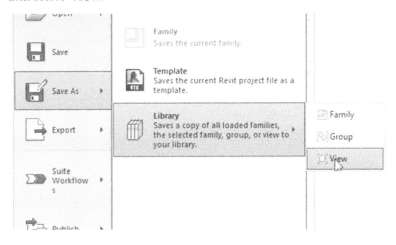

- You will see the following dialog box:

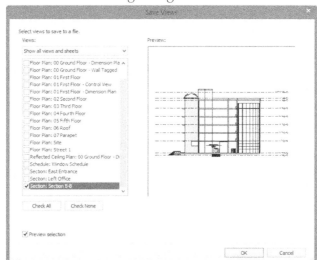

- Save the view(s) in your desired folder. It will be *.rvt.
- For the others to use your view, go to the **Insert** tab, locate the **Load from Library** panel, then click the drop-down list to select **Insert Views from File**:

# EXERCISE 18-2    CREATING DETAILS

**1.** Start Revit 2023.

**2.** Open the file **Exercise 18-2.rvt**.

**3.** Create the following Window Sill Detail using the following data (look at the drawing in the next page first):

    **a.** Create a region using the rectangle tool Solid Black fill (change it to Gray color by duplicating the fill pattern). It should be 200mm (0'-8") width, and roughly 700mm (0'-28") long.

**b.** To its left, create a rectangle Wood1 pattern fill 20mm (3/4") thick, and 300mm (1'-0") long.

**c.** Using a Repeating Detail Component, add Brick to the right of the solid gray rectangle, and attach it to the gray region using the Align command.

**d.** Use Component insert M_Clad Wood Projecting Window-Jamb-Section (Clad Wood Projecting Window-Jamb-Section) and insert it at the top of the two materials, fixing it at the line separating them.

**e.** Use Component insert M_Rough Cut Lumber Section (Rough Cut Lumber Section) and insert it as shown.

**f.** Add Insulation as shown.

**g.** Using Detail Line add lines beside the wooden window to represent the sill as shown.

**h.** Add a Break Line at the bottom. Adjust its size as needed.

**i.** Add text as shown using 2.5mm Arial (3/16" Arial).

**4.** Save the View in a file in your exercise folder and give it the same name.

**5.** Save the file and close it.

**6.** This is the final result:

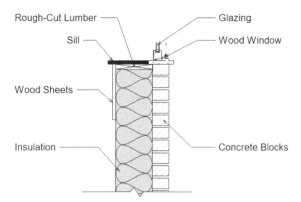

**7.** Create another Draft View, and call it Window Sill Detail – CAD Import.

**8.** Import the AutoCAD file in your exercise folder named Window Sill Detail.dwg, make sure it is Black and White, and leave all the other default values.

**9.** Go to the 00 Ground floor plan and zoom to the office at the right.

**10.** Start the Section command and select the Wall Section. At the context tab, turn on Reference other View and select the Window Sill Detail – CAD Import, and create a section as follows:

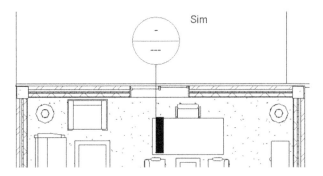

**11.** Double-click the head of the new section; it will take you directly to the CAD import detail view.

**12.** Under the Wall Section category, double-click "Typical Outside Wall."

**13.** The wall and the Masonry layer of the wall, along with floor, have been assigned a keynote code. Check that using Edit Type.

**14.** Using the Keynote Element and Keynote Material, annotate the section to resemble the following:

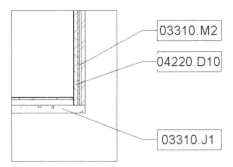

**15.** Save the file and close it.

# NOTES

## CHAPTER REVIEW

1. You can tag in 3D without any restrictions:

   **a.** True

   **b.** False

2. To create a detail in Revit you need to create _____ view.

3. One of the following is false:

   **a.** You can add an Element Keynote

   **b.** You can add a Material Keynote

   **c.** Keynotes for each wall are preset

   **d.** You can add a User Keynote

4. You can import a CAD file and consider it as detail without changing anything:

   **a.** True

   **b.** False

5. To tag all walls in a view in one step, we will use _____ command.

## CHAPTER REVIEW ANSWERS

1. b

3. c

5. Tag All

## CHAPTER 19

# CREATING GROUPS AND REVIT LINKS

## This Chapter Contains

## CREATING GROUPS

- If you ever used AutoCAD, Group in Revit is like Block in AutoCAD.
- Grouping elements in Revit will cut production time significantly if you are using repeating layouts like in designing schools and colleges.
- There are three types of groups in Revit:
  - **Model Groups:** which contain model elements
  - **Detail Groups:** which contain view-specific elements such as text, dimension, tags, and filled regions
  - **Attached Details Group:** which contains both model elements and detail elements together. Revit will not create a group containing both model and view-specific elements, so it will create model group and detail group associated with it.

- There is an associativity inside the group. If you insert it more than once, then you change one instance, all the other instances will change as well.

### How to Create a Group – Method One

- This is the first method. To create a group, do the following steps:
  - Select the desired elements and/or view-specific elements.
  - In the context tab, locate **Create** panel and click **Create Group** button:

  - You will see the following dialog box; type the name of the group. Click the checkbox Open in Group Editor if you want further to edit it:

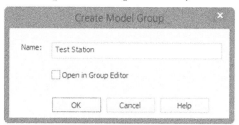

  - When done click OK.
  - Change the group origin by clicking and dragging the origin to a new position, as shown below:

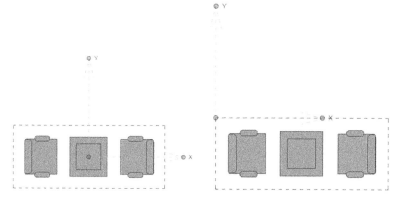

**How to Create a Group – Method Two**

- This is the second method (creating a group without selecting elements).
  - Go to **Architecture** tab, locate **Model** panel, expand the Model Group and click **Create Group**, as shown below:

  - The following dialog box will appear; type in the name of the group and select group type:

  - When the Group Editor opens, you can add elements to the group.

## ADDING GROUPS

- There are two ways to add groups to your project:
  - Using Project Browser
  - From Ribbon

### How to Add Groups from Project Browser

■ All of the groups created in your project will be listed in the Project Browser under the name Groups.

■ Locate your desired group and click-and-drag into the view, as shown:

### Add Groups from Ribbon

■ To add a group from the Ribbon, do the following steps:
  • **For Model Groups:** go to **Architecture** tab, locate **Model** panel, expand **Model Group,** and click **Place Model Group.**

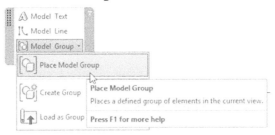

  • **For Detail Groups:** go to **Annotate** tab, locate **Detail** panel, expand **Detail Group**, and click **Place Detail Group**:

  • Using Properties, from Type Selector, select the desired group.
  • You can add multiple copies.
  • Press [Esc] to end the command.

### How to Attach Detail Group to Model Group

- When a model group with an attached detail group is placed in a project, the detail group is not automatically attached. Do the following steps:
  - Select the desired group. In context tab, locate **Group** panel and click **Attached Detail Groups** button:

  - You will see the following dialog box; select the detail group(s) that you to want attach to the model group, as shown, and then click OK:

## MODIFYING GROUPS

- You can perform all modifying commands on a group.
- You can copy or cut them to the clipboard and then paste them anywhere (in the same project, or in another project).
- If you edit an instance of the group, all instances will be affected as well.

- If you want to ungroup an existing group (elements will not be deleted), select it, and from the context tab, click **Ungroup** in the Group panel:

- To rename a group, right-click on the group in the Project Browser and select **Rename.**
- If you want to delete a group from project, do the following steps:
  - Delete all instances of the group in the project.
  - Select the group name in the Project Browser, right-click, and select **Delete**.
- If you want to exclude element(s) from group instances without affecting other instances of the group, do the following steps:
  - Use the [Tab] key to cycle through the group elements up until you select your desired element.
  - You will see the Group Member icon.
  - Click the icon to exclude it from that instance, as shown:

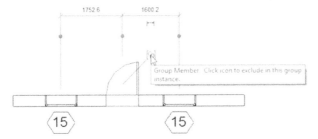

- To replace one group with a different group, do the following steps:
  - Select the group to be replaced.
  - In Properties, using Type selector, select the name of the new group.
- To duplicate groups in the same project, do the following steps:
  - Go to Project Browser.
  - Locate desired group.
  - Right-click and select Duplicate.
  - Or using Properties, select the name of the desired group, click the Edit Type button, and then Duplicate.

## EDITING GROUPS

- Editing here means adding or removing elements from the group; all changes will be reflecting on all instances of the group.
- To edit groups in the group editor, do the following steps:
  - Select one instance of the group.
  - In context tab, locate **Group** panel and click **Edit Group** button.

  - Group Editor will be on, only the group will be clear, everything else, will be dimmed, and you will see the **Edit Group** context tab, which appears as the following:

  - Make changes to the group as needed. You can add, remove, and attach elements, as shown later:
  - When you are done, click **Finish**. All groups with the same name will be updated.

## GROUPS IN OTHER PROJECTS

- You can save your groups as a project file, then use them in other projects.
- To save groups as a file, do the following steps:
  - Using File Menu, expand **Save As**, expand **Library**, and click **Group.**

- In the Save Group dialog box, in the **Group to Save** drop-down list, select a group as shown:

- Pick the desired folder to save the project in, and click the Save button.
- To use the group (which you saved it as a project) in other projects, do the following steps:
  - Open the file in which you want to load the group.
  - Go to **Insert** tab, locate **Load from Library** panel, and click **Load As Group** button:

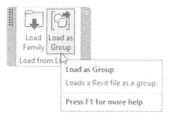

- In the Load File as Group dialog box, go to the folder in which the group resides, and select it.
- Click **Open**. The file is added as a group definition in the current project.
- The group is available now for placing into your project.
- Any Revit project can be loaded into other files as a group.
- Groups can be edited externally from any project, then reloaded again. When you reload a group into a project, the following message will appear if you are loading a file with the same name as an existing group in your project, as shown:

- Click Yes to replace all instances of the group with the new information. If you click No, the group will still be loaded, but with a different name (Revit will add a number after the name).
- Click Cancel to stop the process.

### How to Open a File as a Group in Another Open File

- To open a file as a group in another open file, do the following steps:
  - Open a project file in which you want to place a group.
  - Open the file that you want to use as a group.
  - Go to **Architecture** tab, locate **Model** panel, expand **Model Group** and click **Load as Group into Open Projects** button:

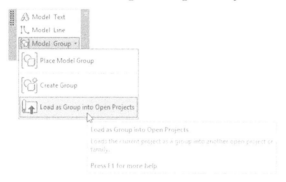

- In the Load into Projects dialog box, select the projects into which you want the currently active project to be loaded as a group.
- Click OK. The group is now available for use in the other projects.

## EXERCISE 19-1    USING GROUPS

**1.** Start Revit 2023.

**2.** Open the file **Exercise 19-1.rvt**.

**3.** You should be in the Gents Room view in floor plans; if not, go to this view.

**4.** Add the following elements (all available in the project):

Start Wall command, change Location Line to Finish Face Interior.
Using wall family Interior – 66 mm partition (Interior - 5 1/2" Partition (1-hr)) draw a 1430 × 1430 (4'-6" × 4'-6") wall.

Add a Single Leaf Door (with tag) 600 × 2000mm (Single-Flush 26" × 76") at the middle of the wall.

Add M_Toilet Commercial Wall-3D -380mm Seat Height (Toilet-Commercial-Wall-3D 15" Seat Height).

5. You should receive the following image:

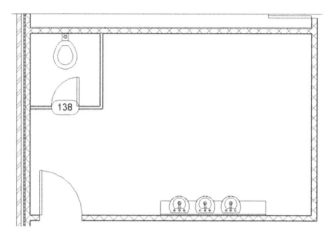

6. Select the two walls, the door with the tag, and the toilet.

7. Create a group, name the Model Group as **Toilet**, and the Attached Detail Group as **Door Tag**.

8. Move the origin point to the upper left corner of the toilet.

9. Using the Place Model Group, from Properties select Toilet group, place a copy of the group directly to the right of the original group (you will notice that door tag is not displayed).

10. Click Finish to complete the process.

11. Select the new group. Using the Attached Detail Groups button, select the Door Tag detail group and click OK. The tag is added to the instance of the group.

12. Using the Project Browser, place two more copies of the model group Toilet in the project.

13. The last group does not need the right wall. Hover over the element that you want to exclude, press [Tab] until it highlights, and select the element.

14. Click the icon to only exclude the wall from that group. Click in empty space to complete the process. But you can see that it did not solve the problem, as the horizontal wall is still too short to meet the vertical wall.

15. Select the last group and Ungroup it. Delete the vertical wall, and extend the horizontal wall.

16. Using the Application Menu, select Save As / Library / Group. In the Save Group dialog box, go to your class folder and save the group with the same name as the group.

17. Save and close the file.

## LINKING MODELS – INTRODUCTION

- If you have ever used AutoCAD, this is like XREF command.
- You can link any RVT file inside another RVT file; if the original changes the linked copy will update accordingly.
- Link the model once, but you can place it as many times as you wish. The linked model can be moved, copied, rotated, and arrayed.
- To place the linked model in the exact place, use Reference Planes (you can lock the linked model to the reference planes if needed).
- You can schedule elements from the linked model in the host project.
- You can tag doors, windows, rooms, and areas of the linked model in the host model.

## LINKING MODELS

- To add a linked model to the host project, do the following steps:
  - Go to **Insert** tab, locate **Link** panel, and click **Link Revit** button as in the following:

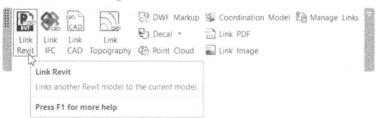

- You will see the Import/Link RVT dialog box; select the file that you want to link. Before opening the file, set the Positioning as shown. Click Open to start the placing process:

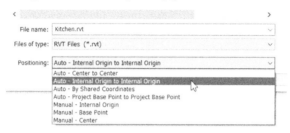

- Go to the Project Browser and locate Revit Links as shown below; if you need more copies of the link, drag and drop into the project:

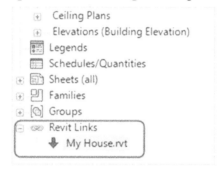

- You can also use the Copy command and Copy to clipboard to copy instances of a linked model. The new copies are also links as well.

## MANAGING LINKS

- To control the linked model, go to **Insert** tab, locate **Link** panel, and click **Manage Links** button:

▪ Another way to reach the same command is when you select a linked file, the context tab will show a panel called Link; click the **Manage Links** button:

▪ In Manage Links dialog box, go to the **Revit** tab, and you can modify other properties of the link:

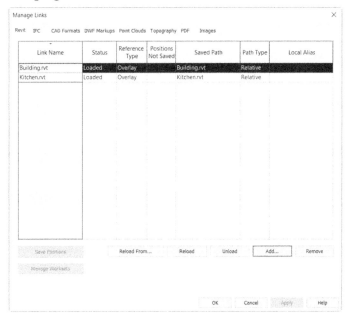

▪ Control all or any of the following:
  • **Status**: shows the current status of the linked file (Loaded or Not Loaded). This is read-only.
  • **Reference Type**: there are two choices, either Attach or Overlay. To understand Reference Type, imagine multiple people using linked files. Let's assume that _**B**_ brings in a linked model to his/her file from _**A**_. Then _**C**_ brings in _**B**_ file. What will _**C**_ see in his/her file? If an

Attachment, **_C_** will see both the **_A_** model and the **_B_** model. If Overlay, **_C_** will see only the **_B_** model.
- **Position Not Saved**: Works in combination with Save Positions. It is active if the link is part of a shared coordinate environment.
- **Saved Path**: Displays the location of the original file.
- **Path Type**: The path type can be set to Relative or Absolute. If absolute, Revit will save the exact path of the linked model, remembering the drive and folder(s), which means if the file is moved to a different place, Revit will report an error. If Relative (the default option), Revit will only save the name of the folder containing the file, ignoring the other superior folders and drive. So, in case of moving your folder to another place, Revit will be able to locate it.
- **Local Alias**: shows an additional name for files linked to Worksets.

- The lower buttons are:
  - **Save Positions**: use this button (if there are shared coordinate systems used in the host model) to save a linked model to a position in the host project.
  - **Reload From**: use this button to reload the file if it was moved to another folder or drive.
  - **Reload**: if you use Unload, use Reload to bring back the file.
  - **Unload**: to unload the file so that it is not shown in the project but is still linked.
  - **Add**: links a Revit model and places an instance in the current view
  - **Remove**: to delete the linked file from the host file.

## VISIBILITY / GRAPHICS OVERRIDES IN LINKED MODELS

- Using the Hide in View command while you have the Host project and the linked project will work fine, if you choose Elements and not Category.
- If you use the Category option, all elements in both the host and linked projects will be hidden.
- But what if you want to change things in the host project without affecting the linked model and vice versa?

### Display Settings in Linked Model

- Use the Visibility / Graphics dialog box to set view options for the linked model.

- To modify Display Settings in a Linked View, do the following steps:
  - Start the Visibility/Graphics Overrides dialog box.
  - Go to **Revit Links** tab. The default in Display Settings is **By Host View**. Click this button to make the desired changes:

  - The **RVT Link Display Settings** dialog box will come up; go to the **Basics** tab and select **By linked view** or **Custom**. Select the **By linked view** option, and then you can select from a list of related linked views in the linked model, as shown:

  - Select Custom, so you can specify each setting individually:

  - If you select Custom, click the Model Categories tab, and you will see everything dimmed, and Model Categories is set to By host view;

change it to Custom, so you can set what is to be shown and what is to be hidden in the linked model:

**NOTE** *In Visibility / Graphics dialog box, you can show the Base point, and Survey point of the linked file which will enable you to move your linked model to the right place easily.*

- If you select the linked model and right-click you will see the following two options:

- Both options will allow you to move the linked model to Project Base point or Internal Origin of the host model

## LINKED MODEL CONVERSION

- If at any point you decide to have the linked file as part of your project, you will lose the connection and any future update; use the Bind command and the linked project will be turned into Group.

- To convert a link to group, do the following steps:
  - Select the link. In context tab, locate **Link** panel, and click **Bind Link** button:

  - The Bind Link Options dialog box will appear. Select the items that you want to include and click OK.

- There are two possible alert messages that may appear:
  - If there is an existing group with the same name
  - If you are bringing elements with families with the same name

### How to Copy Individual Items in a Linked File to the Host File

- You can copy selected elements from the linked model to the host project. Try the following:
  - Hover over the desired element, press [Tab] until it is highlighted, and then Click it.
  - In the context tab, locate the **Clipboard** panel and click **Copy to the Clipboard** button.
  - Click **Paste from Clipboard** to insert the individual element into the project.

## EXERCISE 19-2    LINKING MODELS

1. Start Revit 2023.

2. Open the file **Building.rvt** and save it as **myname_Building.rvt** (example: **Munir_Building.rvt**). Then open the file **Kitchen.rvt**, and save it as **myname_Kitchen.rvt** (example: **Munir_Kitchen.rvt**), then close **myname_Kitchen.rvt** and keep **myname_Building.rvt** open.

3. Link myname_Kitchen.rvt into myname_Building, and place it anywhere initially. Using the Move command, move the kitchen using the upper left corner to the upper left corner of upper room, and then save and close the file.

4. Open the file Exercise 19-2.rvt. The site has four locations for the buildings.

5. Start the Link Revit command, and select myname_Building.rvt. Place initially anywhere. An alert box opens; read it. Simply this is because myname_Kitchen.rvt was inserted as an Overlay, and not attachment.

6. Using the intersection of Grid A & Grid 1, move the building to the upper left corner of one of the four rectangles.

7. Copy the linked model to the other three locations.

8. Start the Visibility/Graphics command. Go to the Annotation Categories tab, uncheck the Grids, and click Apply. All grids in the linked model disappear. Go to the Revit Links tab. Next to the main link, select Halftone.

9. Click the plus sign beside the name of the link to display the four copies. Randomly select one of them, click the Not Overridden button, go to the Basics tab, select the Override display settings for this instance, and set it to By linked view. Click OK twice to finish the command.

10. All of the linked models are set to gray (halftone) with the grids turned off, except for the linked model in which you overrode the Display Settings. It still includes the grids.

11. Save and close the file as **myname_Exercise_19-2.rvt**.

12. Open **myname_Building.rvt**. Select one of the interior walls, right-click, and select Create Similar option and draw several walls inside the building.

13. Using the Manage Links command, go to Revit tab, select the myname_Kitchen.rvt file, and set the Reference Type to **Attachment**.

**14.** Save and close the file. Reopen **myname_Exercise_19-2.rvt**, and the new walls and the kitchen display in all of the linked files.

**15.** Save and close the file.

## COORDINATING ACROSS DISCIPLINES

- Working in Revit with the other two disciplines (Structure and MEP) is inevitable. Elements from the architect model can be copied and/or monitored.
- Structural and MEP engineers can copy column gridlines and levels. This will keep track of any movement or change on the copied or monitored elements.

### Copying and Monitoring Elements

- You can use either:
  - **Copy elements from a linked file**: in this case, Revit will copy the elements to your project, and will keep track of any change; if anything happens, it will inform you to take an action.
  - **Monitor elements from a linked file**: for this part to work, you need to have your own elements (column gridlines, levels, etc.). You will ask Revit to link any element from your project to an equal element from the linked file (no copying will take place here). If the element from the linked project is moved or changed, you will be notified.
- To copy and monitor elements, do the following steps:
  - Go to **Collaborate** tab, locate **Coordinate** panel, expand the **Copy/Monitor** button and click **Select Link** or **Use Current Project.**

- Select the desired link and the following context tab will appear:

- Before you copy and/or monitor elements, you can change the Copy/Monitor options by clicking **Options** button. This will replace any element with a replacement, as shown below:

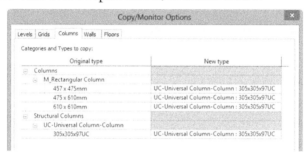

- Using the Tools panel, click the **Copy** button. In the Option Bar, select **Multiple** option. Select all of the desired elements to copy and monitor.
- When done, click the Finish button (not the (✔) symbol).
- Selected elements are now copied into your model. Revit displays a symbol-like break line to designate that the elements are being monitored.
- When you are done, click Finish (✔).
- To monitor elements only, do the following steps:
  - Go to **Collaborate** tab, locate **Coordinate** panel, expand **Copy/Monitor** button, and click **Select Link** or **Use Current Project**.
  - Select the desired link, and then click the **Monitor** button.
  - Click the desired element from your project to monitor, and then select the equivalent element from the linked project.
  - When you are done, click Finish.

**Coordinating Monitored Elements**

- If you did either copy and/or monitor, and the linked file was modified, you will be notified to do a Coordination Review.

- If you open a project containing a link with copied or monitored elements which was altered, the following warning message will be displayed telling you that a coordination check should happen:

- To do a coordination review, do the following steps:
    - Go to the **Collaborate** tab, locate the **Coordinate** panel, expand the **Coordination Review**, and click **Select Link** or **Use Current Project**:

    - Select the desired link. The Coordination Review dialog box appears showing the modifications as shown:

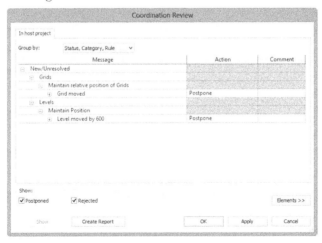

- If you select an element in the list and click the **Show** button, Revit will take you to the model to show you the issue to be reviewed in this element.
- To solve the problem, select an **Action** next to each of the elements as shown below. You may add a comment to justify your action:

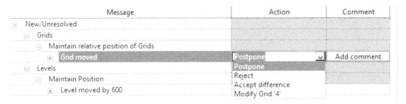

- To create an HTML report, click the Create Report button.
- When done, click OK.

### Interference Checking

- Interference Check examines if interferences took place between different types of elements. For example, there might be a door obstructing a column.
- To run an interference check, do the following steps:
    - Go to the **Collaborate** tab, locate the **Coordinate** panel, expand **Interference Check** and click the **Run Interference Check** button:

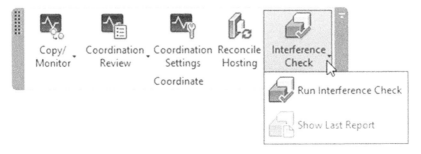

- The following dialog box will appear; select the Categories from each project that you want to check, as shown below. You can select from the Current Project or a linked file if applicable:

- When done click OK. The Interference Report dialog box opens as shown below:

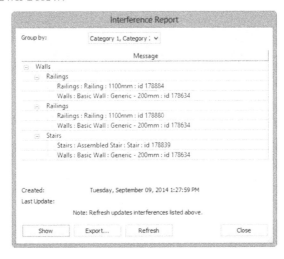

- Use the buttons to display or export a report as needed.
- Close the dialog box. Try the needed changes in the project(s).
- To ensure that all of the interference was cleared out, do the following steps:
  - Go to **Collaborate** tab, locate **Coordination** panel, expand **Interference Check** and click **Show Last Report** button:

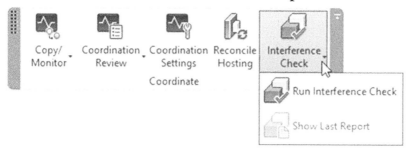

- In the Interference Report dialog box, click the **Refresh** button.
- If the interference has been cleared, the dialog box will remove the interference from the report.
- When done, close the dialog box.

## LINKING OTHER FILE TYPES

- You link other file formats like:
  - Link IFC (IFC file extension is an **I**ndustry **F**oundation **C**lasses file.) It is used by Building Information Modeling (BIM) programs to hold models and designs of facilities and buildings.
  - Link Topography, which was produced by Autodesk Civil 3D software
  - Link PDF
  - Link Image
- All of these buttons exist in the Link panel, in Insert tab:

- Almost everything applies to Linking Revit Model, applies here, so no need to discuss it again.

## EXERCISE 19-3    COORDINATING ACROSS DISCIPLINES

1. Start Revit Architecture 2023.

2. Open the file named Modern_Commercial.rvt, and save it as myname_ Modern_Commercial.rvt, and close it.

3. Open Exercise 19-3.rvt, and save it as myname_Exercise 19-3.rvt.

4. Using the Link Revit command, link the project myname_Modern_Commercial.rvt. Set Positioning to Auto - Center to Center.

5. Start Copy/ Monitor command, click Select Link, and then select the linked file.

6. Using the Copy/Monitor tab and Tools panel, click Copy. In the Options Bar, select Multiple. Hold down [Ctrl] and select all of the grid lines. In the Options Bar, click Finish and press [Esc] twice.

7. The elements have been physically copied into your project.

8. We are still in Copy / Monitor command, using the Copy/Monitor tab, Tools panel, click Options button. In the Copy/Monitor Options dialog box, in the Columns tab, locate Structural Columns, M_Concrete-Square-Column, Size 300 × 300 (Concrete-Square-Column, Size 12 × 12) (this is the size used in the linked model), in the New Type, select Copy original Type, then click OK.

9. In the Tools panel, click Copy; don't select Multiple. Click on one of the architectural columns. You are copying the same column to your current project. Continue copying columns. When you are done, click Finish.

10. Hide the linked model in order to see the grid lines and columns.

11. Go to the South elevation view, and you will see levels from the linked model, and two levels (Level 1, and Level 2) from your model. Stretch the two levels of yours outside, so you can see them clearly.

12. Start Copy / Monitor, and select the linked project.

13. Click the Monitor button, click Level 1, and then click 00 Ground. This way you will monitor the 00 Ground level, without physically copying it to your model.

**14.** Click the Options button; in Level tab, set Offset level = 400 (1'-4"), and Prefix = *My-*, click OK. Make sure you are selecting the Copy button, and click 01 First level.

**15.** Continue copying the other levels using the same way. When done click Finish. Hide the linked model to see what you copied.

**16.** Save myname_Exercise 19-3.rvt, and close it.

**17.** Open myname_Modern_Commercial.rvt.

**18.** Go to the 00 Ground floor plan view.

**19.** Move the F gridline 1000mm (3'-4") to the left.

**20.** Save it, and close it.

**21.** Open myname_Exercise 19-3.rvt. Revit will tell you right away that you need to make a Coordination Review because the linked model has changed.

**22.** Go to Level 1, Unhide the linked file, and you can see two F gridlines.

**23.** Start the Coordination Review command, and select the linked model.

**24.** Select Modify Grid F to accept the change from the linked model.

**25.** Once you click the OK button, you will see only one grid line F, which is like the linked model.

**26.** Save and close the file.

## NOTES

## CHAPTER REVIEW

1. Any Revit project can be loaded into other files as a group:

   **a.** True

   **b.** False

2. If you have a Revit project linked into your project, to keep it with you and convert it to group, use _____ command.

3. The following statements are true except one:

   **a.** You can ungroup any group.

   **b.** You can select one element of the group and hide it.

   **c.** You can save the group as an RVT file.

   **d.** You cannot load the group into my current project.

4. Copy or Monitor are the same command:

   **a.** True

   **b.** False

5. For linking files:

   **a.** You can copy elements from linked to the host.

   **b.** While you are copying elements from linked to host, you can't change them.

   **c.** If you want to use the Monitor command, you have to have elements of your own.

   **d.** If you copy, or monitor, and if anything changes in the linked file, Revit will let me know.

6. The Interference Check command exists in _____ tab.

## CHAPTER REVIEW ANSWERS

**1.** a

**3.** d

**5.** b

# 20

# *IMPORTING AND EXPORTING FILES IN REVIT*

## This Chapter Contains

- Importing Vector files
- Modifying Imported files
- Importing Raster Image files and PDF files
- Exporting files

## IMPORTING/LINKING CAD FILES

- There are many companies that still use AutoCAD, or use it partially, or have legacy drawings. Revit can make our lives easier by importing these files to its environment.
- Revit can import several types of CAD (vector-based) files:
  - AutoCAD (DWG and DXF)
  - MicroStation (DGN)
  - SAT (the native file type of the ACIS 3D modeling)
  - SketchUp (SKP) into mass and in-place families (you should be in 3D view as a first step)
- You can use either Link (just like we did with Revit files) or import (just like we did in Chapter 3)

**How to Import or Link a Vector File**

- To import a vector file, go to **Insert** tab, locate **Import** panel, and then click **Import CAD** button:

- To link a vector file, go to **Insert** tab, locate **Link** panel, and then click **Link CAD** button:

- Fill out the Import CAD (or Link CAD) dialog box. The top part of the dialog box holds the standard select file options. Please check Chapter 3 for importing options.
- Only the current drawing is imported, not any reference files. You need to bind the reference files to the main file before it is imported.
- Text only appears if you are importing to the current view.
- Use the CAD file to trace over it and snap to its points.
- You can print a mixed drawing—part Revit project and part imported drawing.

**Importing Lineweight**

- One other significant setting for imported drawings is the lineweight. Typically, each color in AutoCAD is associated with a lineweight. Therefore, Revit imports them by color.

- To import lineweights, do the following steps:
  - Before you import a CAD file, go to the **Insert** tab, locate the **Import** panel, and click **Import Line Weights** as shown below:

  - The following dialog box will appear; load a text file that holds the relationships or type them in the dialog box. You can then save them for later use:

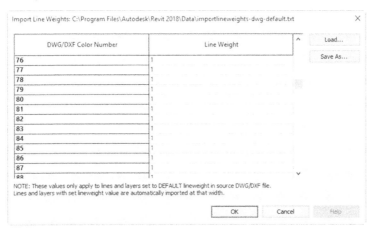

  - Click OK and then import the CAD file.

## MANIPULATING IMPORTED CAD FILES

### Query Imported CAD Files

- Query command is to help you know the type of object in the imported CAD file along with its layers. It also enables you to delete or hide the layer.

- To query imported files, do the following steps:
  - Select the imported file that you want to query. In the Import Instance panel, click **Query** button:

- Click the desired object in the CAD file. You will see the following dialog box displays information about the objects, as shown:

  - If you want to delete it, click **Delete** button.
  - If you want to hide it, click **Hide in view** button.
  - If you want just to see the information, click OK.
- You are still in Query command and can select another object or press [Esc] to end the command.

### Deleting Layers

- If you are creating a file containing both Revit and CAD objects and you want to dispose of some of the layers in the CAD file, without exploding the imported file. Use Delete Layers command. Only objects in the selected layers will be removed.
- To delete layers in the imported files, do the following steps:
  - Select the imported file.
  - In the Import Instance panel, click **Delete Layers** button.

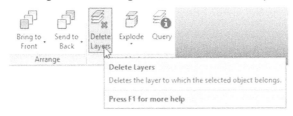

- You will see the following dialog box which displays all layers or levels in the drawing, as shown below:

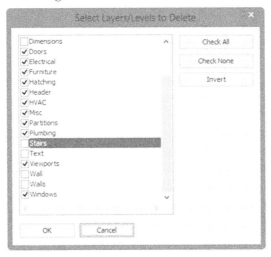

- Select the layers that you want to delete from the imported file and click OK.

## Exploding Imported Files

▪ You can explode the file to convert it to lines, text, curves, and filled regions. Select the CAD file and in the context tab you will see the following:

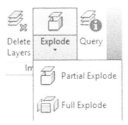

- **Partial Explode:** Explodes all top-level objects but does not explode reference files or blocks in the imported drawing.
- **Full Explode:** explodes everything to Revit objects.

▪ You cannot explode linked files.

### Modifying the Visibility of Imported Files

■ Revit gives you the ability to hide the whole CAD file or only some layers.
■ To hide individual layers, do the following steps:
  • Start **Visibility/Graphics** command.
  • Go to the **Imported Categories** tab. You will see a list for each imported CAD file, and its layers, as shown in the following:

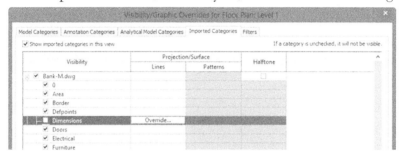

  • Click the plus sign beside the name to see a list of layers in that file.
  • Uncheck individual layers that you do not want to display.
  • Close the dialog box.
■ To turn off the entire file, clear the check mark next to the filename.
■ You can also use Hide in View and Override Graphics in View in the View Graphics panel or in the shortcut menu to modify the view graphics of the CAD file.

## IMPORTING RASTER IMAGE FILES

■ Maybe you need a logo to be used in a title block, or maybe you want to add a raster image to act as a background view or as part of the final drawing.
■ Imported images are placed behind model objects and annotation.
■ To import a raster image, do the following steps:
  • Go to **Insert** tab, locate **Import** panel, and click **Image** button:

- In the Open dialog box, select the desired image file. You can insert bmp, jpg, jpeg, png, and tif files.
- When you click Open, four blue dots and an "X" illustrate the default size of the image file, as shown in the following illustration. Click on the screen to place the image:

- Resize the graphic as needed.

## Editing Raster Files

- Select an image to make changes. Once it is selected, you can resize the image as you did when you first inserted it.
- Select the Lock Proportions option on the Options Bar to ensure that the length and width resize proportionally to each other when you adjust the size of an image.
- Use Modify commands like: Move, Copy, Rotate, Mirror, Array, and Scale images.
- Use Bring to Front and Send to Back to move images to the front or back of other images or Revit elements.
- You can snap to edges of images, as shown:

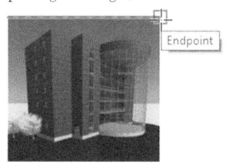

- Go to **Insert** tab. Locate **Link** panel and click **Manage Links** button:

- You will see the following dialog box. Select **Image** tab, you will see the following dialog box, which is the same dialog box discussed previously:

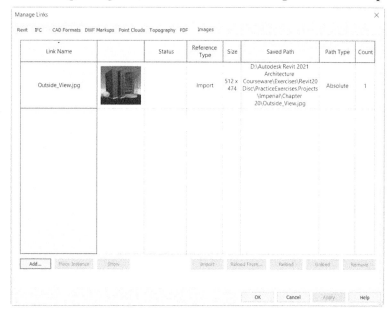

- If you try to remove one of the images, you will see the following warning:

## IMPORTING PDF FILES

- You can import PDF files to your Revit model as raster images
- This feature will help you to combine multiple file format in the same Revit model, since PDF file format is the most used file format
- The PDF file origin can be from any other software

■ To import PDF file, do the following steps:
  • Go to Insert tab, locate Import panel, and click the PDF button:

  • Select the folder, and the file name, click OK
  • You will see the following dialog box:

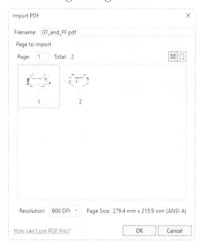

  • If the PDF file contains more than one page, select the desired page, then select the Resolution, and click OK
  • You will see the following:

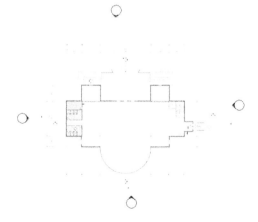

- If you select the PDF instance, you will see the following in the context tab:

- Use Bring to Front and Send to Back, to move PDF file to the front or back of other images or Revit elements
- Use Enable Snaps button to turn on Snapping to the PDF file
- Use the Manage Links dialog box to control the PDF in your model
- All functions discussed for image control in Revit model can be used as well for PDF file imports

## EXERCISE 20-1  WORKING WITH IMPORTED FILES

**1.** Start Revit 2023.

**2.** Open the file **Exercise 20-1.rvt**.

**3.** You are in the Level 1 floor plan view.

**4.** Start the Import CAD command.

**5.** In the Import CAD dialog box, select the AutoCAD drawing file **Ground_Floor.dwg** from your class folder and set the following options:

   **a.** Select Current View Only

   **b.** Colors: Black and White

   **c.** Layers: All

   **d.** Import Units: Auto-Detect

   **e.** Positioning: Auto-Center to Center

**6.** Go to the South elevation view. You can't see anything because the imported DWG is 2D only.

**7.** Go back to the Level 1 floor plan view.

**8.** Select the imported file. Then click Query button.

**9.** Click the arc of one of the doors? Does Revit know that this is a block? If yes, write down the name of the block and the name of the layer:

_____

**10.** Click OK.

**11.** Still in Query command, click one of the text. What is the name of the layer the text objects reside in? _____,

**12.** If you click Delete, all text will be deleted along with the door. Click Delete.

**13.** Start the Visibility/Graphic Overrides dialog box. Switch to the Imported Categories tab. Click the "+" next to Ground_Floor.dwg to see the layers. Turn off the layers Centerline, Elevator, Elevator_Door, and Hatch. Click OK.

**14.** Select the file. In the Import Instance panel, click Partial Explode. The file is exploded. Move the cursor over elements in the project; they became Revit Detail Lines.

**15.** Using Crop Region, create a region almost as big as the imported CAD file (make sure to hide the Crop Region later).

**16.** Go to the Sheets (all) and double-click the existing sheet.

**17.** Drag and drop the Level 1 plan view onto the sheet, leaving room for other information.

**18.** Using the Image command, select the image file called Outside_View.jpg (in your class folder) and place it on the sheet.

**19.** Move and resize the image and insert it in suitable place.

**20.** Go to Level 2 floor plan view

**21.** Start PDF command, to import a PDF file GF_and_FF.pdf (from your exercise folder), and select the first page, with 600 DPI resolution and insert it at the middle of the screen

**22.** Turn on Enable Snaps button

**23.** Draw two detail lines over the grid lines C & D, then put a dimension between the two lines. The dimension should read 6498mm (20'-5.75"). The real distance is 20000mm (65'-7.5")

**24.** Click the PDF file, from Properties, set the horizontal scale = 307.787 (307.629)

**25.** The PDF file now is to-scale, and you can use it to trace it with real Revit walls, doors, etc.

**26.** Save and close the project.

## EXPORTING FILES

- Revit can export to CAD formats, PDF, DWF/DWFx, FBX, gbXML, IFC, ODBC databases, Images and Animations, and Reports:

- You can export all 2D views (e.g., floor plan, ceiling plan, elevations, sections, etc.) and 3D views (whether parallel or perspective).
- Only 3D views export the entire building model; other views create 2D files.

### Exporting CAD Format Files

- You can export views or sheets to DWG, DXF, DGN, and SAT files. To export a CAD format file, do the following steps:
  - If you are exporting only one view, open the view you want to export. If you are exporting the model, open the 3D view.
  - Go to **File** Menu, select **Export**, then select **CAD Formats** option. You will see the following dialog box:

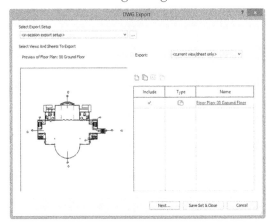

- If you have an existing export setup, you can select it from the drop-down list as shown below:

- To export only the active view, select <Current View/ Sheet Only>.
- To export any views or sheets that are open in this session, select <In session view/sheet set>.
- To export a pre-defined set of views or sheets, select the name from the list if it is available.
- When everything is fine, click the **Next** button.
- Specify the desired folder location and name. If you are exporting to DWG or DXF, select the version in the Files of type drop-down list.
- Click OK.

## How to Create an Export Setup

- To create an Export Setup, do the following steps:
  - In the DWG Export dialog box, below the Select Export Setup list, click Modify Export Setup (the button with the three dots).
  - You will see the following dialog box which contains all CAD objects you can export. For layers, select an existing Layer standard available in the software, as shown below, or create a new one:

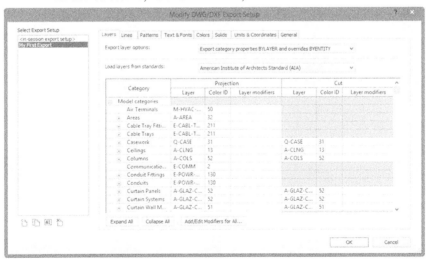

  - Go to Lines, Patterns, and Text & Fonts tabs to pick how these things will be exported.

- Go to the Colors tab to select how to export colors, using either Index colors (255 colors) or True color (RGB values).
- Go to the Solids tab (available only if 3D views are selected), to select how to export 3D elements; either Polymesh or ACIS solids.
- Go to the Units & Coordinates tab to specify what DWG the unit is (pick one of the following: Foot, Inch, m, cm, or mm), and what is the coordinate system basis (pick from the following: internal or shared).
- Finally go to the General tab to set up how rooms and room boundaries are exported, if there are any non-plottable layers, and the DWG file format.

### How to Create a New Set of Views/Sheets to Export

- To create a new set of views or sheets to export, do the following steps:
  - Start the appropriate Export CAD Formats command.
  - In the Export CAD Formats dialog box, click the **New Set** button:

  - In the New Set dialog box, type a name and click OK:

  - You will see a list of 2D and 3D views, as shown in the following:

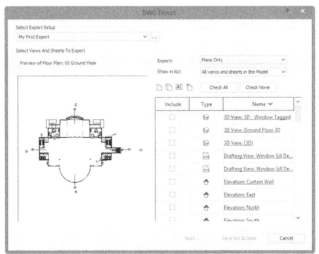

- If you have a huge list of views, use the Show in List drop-down to downsize the list.
- Select the views and/or sheets that you want to export from the project. Use Check All or Check none to help you speed up the process of selection. When you are done, click Next to complete the export process.

### Exporting PDF Files

- You can export views or sheets to PDF. Do the following steps:
  - Go to **File** Menu, select **Export** then select **PDF** option. You will see the following dialog box:

- Specify the Export Range; either the Current Window (View), the Visible potion of the current window, or Selected views/sheets (if you pick the third choice, you will see a dialog box to select the desired views or sheets).

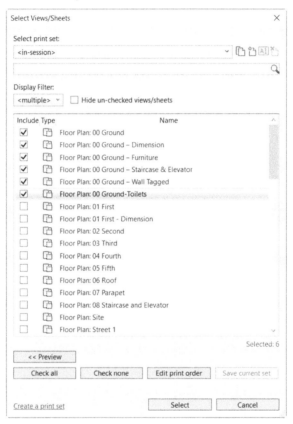

- In this dialog box, you can click Preview, to preview each sheet/view, Check all, Check none, and Edit print order. If you click Edit print order you will see the following dialog box:

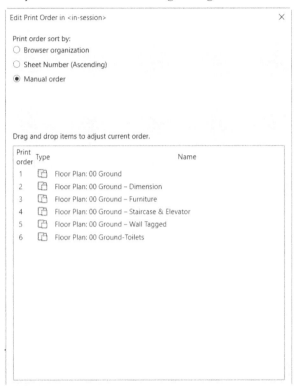

- Use this dialog box to set the print order of your sheets/views. Use Manual order to set your own order method
- After finishing the sheets/view selection type in the name of the PDF file, and whether the views/sheets will be combined in single file or each one in a separate file, then specify where you want to save the file
- Specify the Page Size, and the zoom level
- Specify the Paper Placement
- Specify the Orientation (Landscape, or Portrait), the Appearance, and whether Vector, or Raster Processing
- Specify the other Options and click Export
- Click OK

## EXERCISE 20-2   EXPORTING FILES

1. Start Revit 2023.

2. Open the file **Exercise 20-2.rvt**.

3. Go to the File Menu and select Export / CAD Formats / DWG.

4. Click New Set, and name it My_Set.

5. Using Show in list, pick Views in the Model, and select only Floor Plan: 00 Ground, and Floor Plan: 01 First.

6. In the Select Export Setup area, click Modify Export Setup (the small button with the three dots).

7. Go to the Layers tab and load layers from the American Institute of Architects Standard (AIA).

8. Go to the Text & Fonts tab for Text behavior when exported, select Preserve Editability. Scroll down in the list of fonts and map Arial to Calibri.

9. Go to the General tab, select Export rooms and areas as polylines, and Click OK.

10. Click Next. In the Export CAD Formats - Save to Target Folder dialog box, set the Save In: folder to the class folder. Set the Files of Type: to AutoCAD 2013 DWG files.

11. Set the Naming to Automatic-Long (Specify prefix), do not change the name, and then click OK. Revit generates DWG files for each selected view using the setup you defined.

12. Go to the 3D view.

13. Start the Export command again and select CAD Formats / DWG Files.

14. Set Export to <Current View/Sheet only>, click Next, do not change the name of the file, and then click OK.

15. Using AutoCAD (2013 or up), open the 2D and 3D files to check the exported geometry or view them in Windows Explorer.

16. Go to the File Menu and select Export / PDF

17. Select the third choice in Export Range, click the pencil button at the right, and specify all the Sheets only. Don't Save, click Select

**18.** Do not change the name, and leave the checkbox of Combining turn on, and Location your exercise folder

**19.** For Size, set Zoom = 100%

**20.** Select the Vector Processing

**21.** Leave all the other options as is, click Export

**22.** View the resultant PDF file in PDF viewer

**23.** Close the file.

# NOTES

## CHAPTER REVIEW

1. _____ command is to help you know the type of object in the imported CAD file along with its layers.

2. You can insert the same Raster file into your project more than once:

    **a.** True

    **b.** False

3. You cannot import to my Revit project:

    **a.** DWG and DXF

    **b.** SKP file

    **c.** XLSX file

    **d.** SAT file

4. All Revit views exported will be included in one DWG file:

    **a.** True

    **b.** False

5. While exporting Revit files to AutoCAD files, which statement is false:

    **a.** You can control which layers to be exported

    **b.** You can control how linetypes are exported

    **c.** You can control how hatch patterns are exported

    **d.** You can control how colors are exported

6. When you import AutoCAD files into your Revit project, there are two types of explode you can perform, they are _____ and _____.

## CHAPTER REVIEW ANSWERS

1. Query

3. c

5. a

## CHAPTER 21

# CREATING MASSES IN REVIT

## This Chapter Contains

- Introduction to Massing Studies
- Placing Mass Elements
- Creating Conceptual Massing
- Setting Work Planes
- Creating Mass Forms
- Dynamic Editing for Conceptual Massing
- Working with Profiles and Edges
- From Massing to Building

## INTRODUCTION

- One way to create your building is to create a mass, then convert the different faces to walls, floors, roofs, and curtain systems.
- You can accomplish this by using two commands in Revit:
  - **Place Mass command**, in which you will use pre-made 3D shapes with adjustable dimensions
  - **In-Place Mass**, in which you will create more complex shapes, but it will be only for this project

- In general, mass elements will not be displayed by default in projects. In order to show them, go to **Massing & Site** tab, locate **Conceptual Mass** panel, and click **Show Mass Form and Floors** as shown below:

- If you want to use the first option: Show Mass by View Settings, go to the Visibility/Graphic Overrides dialog box, go to Model Categories, and turn on Mass option, as shown below:

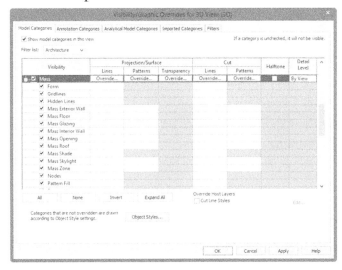

- *Even if you forgot to turn on the visibility, once you started the Place Mass command, Revit will show you the following message:*

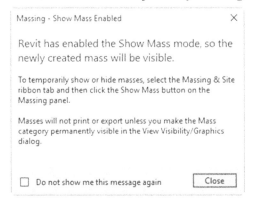

**NOTE**

- *SketchUp files can be imported into Revit Mass families and then you can use them in Revit like other massing elements.*

## PLACE MASS COMMAND

- These are the pre-made shapes that are included in Revit (found in Mass folder in libraries).
- They include shapes such as: Box, Cylinder, Cone, Dome, Gable, and so on.

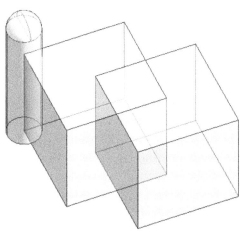

- To place a mass element, try the following:
  - Go to the **Massing & Site** tab, locate the **Conceptual Mass** panel, and click the **Place Mass** button:

  - Normally there will be no mass family loaded into your template file, so Revit will ask you to load a family:

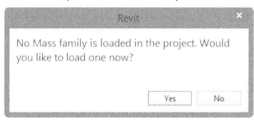

  - Using the Type Selector, select a mass type, as shown below:

  - Using the context tab, locate the **Placement** panel, select either **Place on Face** or **Place on Work Plane**, and then place the mass in your model:

- Using the Option bar, click the checkbox **Rotate after placement** on to allow you to perform the rotation command after the placement, or simply use the spacebar to do that (in case of the spacebar, only 90° is permitted).
- When done press [Esc] twice.

■ To modify the size of the mass, use one of the following ways:

- Click the mass to show shape handles, then move them to change the dimension of the mass, as shown below. This way may not be the best way if you make precise changes:

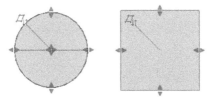

- Click the mass and make changes in the Properties, as shown below:

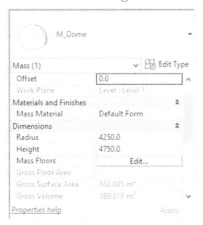

■ You can move a mass to a different work plane. Do the following steps:

- Select the desired mass.
- Using the context tab, locate the **Work Plane** panel, and click the **Edit Work Place** button:

- You will see the following dialog box; identify the new work plane:

- You can change the material of the mass if you do not like the transparent material assigned to it.
- Go to **Modify** tab, locate **Geometry** panel, click **Paint** button, and then add materials to selected faces of a mass:

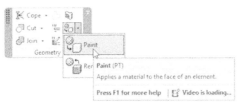

- To set material for all mass elements in a project, do the following steps:
  - Go to **Manage** tab, locate **Settings** panel, and click **Object Styles** button:

  - Click the plus sign next to **Mass** category, and then select the material for the Form as shown here:

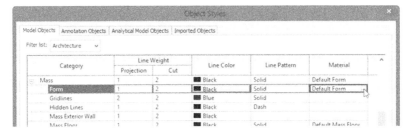

- Click in the Material column and then click the small button (with the three dots); the Material dialog box will appear.
- Using the Material dialog box, select the desired material to be used for all masses in your project.
- Click OK to close the dialog boxes.

■ *After placing masses in Revit you can:*

**NOTE**
- *Use the Join command, to join two masses or more*
- *Use the Cut command, to subtract masses from each other (always select the big one first)*

## EXERCISE 21-1   BASIC MASS ELEMENTS

**1.** Start Revit 2023.

**2.** Open the file **Exercise 21-1.rvt**.

**3.** Turn on Show Mass Form and Floors.

**4.** Start Place Mass command and load the Box and Cylinder.

**5.** Using the lines in Level 1, place two boxes using Height = 8000 (Height = 20'-0") aligned with the outside lines, and the two cylinders using Radius = 2500 (Radius=8'-4"), and using Height = 4000 (10'-0").

**6.** Look at your model in 3D.

**7.** Join the two boxes, and cut the two cylinders.

**8.** Go back to Level 1.

**9.** Add two boxes using the inside lines, using Height = 8000 (Height = 20'-0").

**10.** Select both new boxes using Properties, set the Offset=8000 (Offset=20-0") to ensure they will lie above the first two.

**11.** Join the new boxes together and then join the lower to the upper.

**12.** Using the Paint command, select Brick, Common, and assign all the faces of the two lower boxes with this material.

**13.** Assign Aluminum to the faces of the upper boxes.

**14.** Assign Concrete, Cast-in-Situ (Concrete, Cast-in-Place gray) to the other faces.

**15.** Save and close the file.

## IN-PLACE MASS & CONCEPTUAL MASS

- In order to create more complex shapes, use one of the following two methods:
  - **In-Place Mass**, which will be created only in your current project.
  - **Conceptual Mass**, which means you will open a new file using the Conceptual mass template file.
- To create a mass using the in-place mass, do the following steps:
  - Go to **Massing & Site** tab, locate **Conceptual Mass** panel, and click **In-Place Mass** button:

- You will see the following dialog box; type a new name for the mass:

- You will see the Conceptual Mass Environment Ribbon, as shown here:

- To create masses using massing family, do the following steps:
  - Go to File Menu, expand **New** and click **Conceptual Mass**.
  - A separate interface will start, as shown below. It resembles the In-Place ribbon, with two differences; the **Load into Project**, and **Load into Project and Close** buttons.

- The Conceptual Mass family template contains one level and two reference planes to help you draw the profile, as shown below:

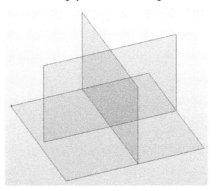

## SETTING WORK PLANES

- Your first step to create the profile(s), (which later will be converted to a 3D model) is to create a work plane, and there are three ways to do that:
  - Use existing faces of existing masses as work planes
  - Use levels as work planes
  - Create and use reference planes as work planes
- To see the workplane in your view, locate the **Work Plane** panel, and then click the **Show** button:

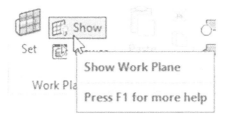

- To establish a work plane by face, do the following steps:
  - Go to **Create** tab, locate **Draw** panel, and select one of the drawing tools.
  - In context tab, locate **Draw** panel, and click **Draw on Face** button.

- Hover your mouse over the desired face of an existing mass; the face will highlight and you can then draw the sketch, as shown below:

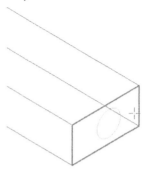

- You can use other faces to sketch more profiles.
▪ To establish a work plane using levels or reference planes, do the following steps:
  - Go to **Create** tab, locate **Draw** panel, and select a drawing tool.
  - In context tab, locate **Draw** panel and click on **Draw on Work Plane** button.
  - In the Options Bar, in Placement Plane drop-down list, select a Level or named Reference Plane, as shown below:

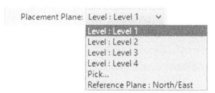

  - Draw the sketch on the plane, as shown below:

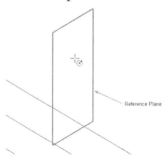

  - If you are in a plan view, the associated level will be considered the current work plane.

- To change the work plane of a sketch, do the following steps:
  - Select the sketch
  - In the Options Bar, click the **Show Host** button:

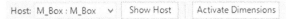

  - The host work plane will be highlighted. Using the Host drop-down list, pick a new host. The sketch moves to the selected plane.

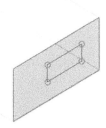

## CREATING MASS FORMS

- You can create six types of forms, they are:
  - Extrusion
  - Sweep
  - Blend
  - Swept Blend
  - Revolve
  - Loft
- These commands are not listed in the Ribbon. Revit will use the suitable command based on the 2D sketch you created.
- There are two commands, either create a solid form or a void form.

### Creating an Extrusion

- Draw a closed or open 2D object as a profile:

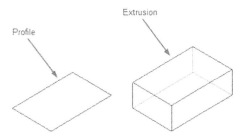

- To create an extrusion, do the following steps:
  - Using Draw panel, pick a drawing tool and select the right work plane.
  - Create a closed profile (if you want a solid) or open profile (if you want a surface).
  - Select the profile.
  - In the context tab, locate **Form** panel, click **Create Form**, and then **Solid Form**. Revit will extrude a solid/surface from the profile
- For some shapes, like circles, Revit will ask you if you want a cylinder or sphere:

### Creating a Revolve

- Draw an axis plane and a profile that are both in the same plane:

- To create a Revolve, do the following steps:
  - Using Draw panel, pick a drawing tool and select the right work plane.
  - Create a closed profile (if you want a solid) or open profile (if you want a surface).
  - Create a line in the same reference plane or level as your axis.
  - Select both the axis line and the profile.
  - In the context tab, locate **Form** panel, click **Create Form**, and then **Solid Form**. Revit will revolve a solid/surface from the profile.
  - Select the entire form to set the Start and End angle using Properties, as shown here:

| Form (1) | | |
|---|---|---|
| Constraints | | |
| Start Angle | 0.000° | |
| End Angle | 360.000° | |
| Graphics | | |
| Visible | ✓ | |
| Visibility/Graphics O... | Edit... | |
| Materials and Finishes | | |
| Material | <By Category> | |
| Identity Data | | |
| Subcategory | Form | |
| Solid/Void | Solid | |
| Properties help | Apply | |

### Creating a Sweep

■ Draw a path and a profile, as shown below. A Point element will be used here to locate the profile on the path:

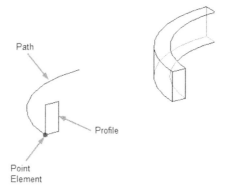

■ To create a sweep, do the following steps:
  • Using Draw panel, pick a drawing tool and select the right work plane.
  • Sketch a path for the sweep.
  • Click Point and specify its location on the path.
  • Go to the 3D view.
  • Pick the Point, a reference plane perpendicular to the path will be highlighted, and it will be the plane to create the profile.
  • Sketch the desired profile on the new reference plane.
  • Select both the path and the profile.
  • In the context tab, locate **Form** panel, click **Create Form**, and then **Solid Form**. Revit will sweep the profile along the path.

### Creating a Blend

■ Draw two profiles in different planes, as shown in the following:

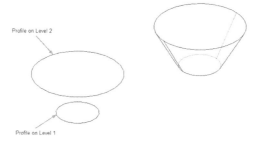

- To create a blend, do the following steps:
  - Using Draw panel, pick a drawing tool and select the right work plane.
  - Sketch a closed profile. Select another level or reference plane, and create a second closed profile.
  - If the second profile is open, Revit will create a form by blending the closed and open profiles.
  - If both profiles are open, Revit creates a surface instead of a solid.
  - Select both closed profiles.
  - In the context tab, locate **Form** panel, click **Create Form**, then **Solid Form**. Revit will blend the two profiles.

### Creating a Swept Blend

- Draw two profiles in different planes connected by a path, as shown here:

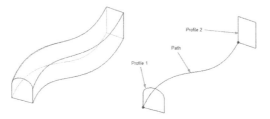

- To create a Swept Blend, do the following steps:
  - Using Draw panel, pick a drawing tool and select the right work plane.
  - Sketch the path.
  - Place two Points along the path, or within it.
  - Select the first Point. In a 3D view, a reference plane perpendicular to the path is highlighted. Use the drawing tools to sketch the first profile.
  - Select the second Point Element. A reference plane perpendicular to the path is displayed. Use the drawing tools to sketch a second profile.
  - Select the path and the two closed profiles.
  - In the context tab, locate **Form** panel, click **Create Form**, and then **Solid Form**. Revit will create a swept blend from the two profiles and the path.

## Creating a Loft

- Draw more than two profiles in different planes, as shown in the following images:

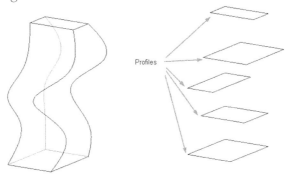

Profiles

- To create a loft, do the following steps:
  - Using Draw panel, pick a drawing tool and select the right work plane.
  - Sketch a closed profile.
  - Specify the workplane where you want to draw the second profile.
  - Use the drawing tools to sketch a closed profile.
  - Continue sketching profiles on different work planes.
  - Select all of the profiles.
  - In the context tab, locate **Form** panel, click **Create Form**, then **Solid Form**. Revit will create a loft out of all profiles.

## Void Forms

- You will use the same methods to create a void, bearing in mind that the void form needs to intersect with the solid form to subtract the volume as shown in the following illustration:

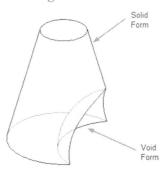

Solid Form

Void Form

- To create a void form, do the following steps:
  - Create the necessary sketches.

- In the context tab, locate **Form** panel, click Create Form, then **Void Form**. Revit will create a void out of the sketches.
- The void automatically subtracts volume out of the solid form it intersects.

## MANIPULATING CONCEPTUAL MASS

- While you are in In-Place editor, you can select either a vertex, edge, or face, and move them to create a more complex shape:

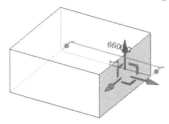

- 3D Control is tripod which will be displayed once the selection takes place (note you have to be in 3D view to see the effect of your work).
- Select either a face, edge, or vertex, as shown here. You will see the 3D control is displayed:

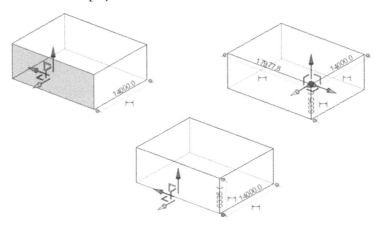

- Hover your mouse over one of the three arrows, until it highlights, then click and drag to make the movement you want.
- To move a face, edge, or vertex along a plane (which is the semi rectangle shape between two arrows) hover your mouse over the semi rectangular shape until it highlights, and then click and drag to make the movement.

- Once you select a face or an edge, you can use the **Rotate** command from the Modify panel.
- You can delete a face, an edge, or vertex, by selecting it, then pressing [Del] key on the keyboard, or clicking **Delete** command from the Modify panel.

## X-RAY – ADDING PROFILES AND EDGES

- By default, Revit will show you only the outside vertices and edges.
- X-Ray mode will show your In-Place mass as edges, profiles, and vertices. To show X-Ray for a mass do the following steps:
  - Select the desired form.
  - In context tab, locate **Form Element** panel, and click X-Ray button:

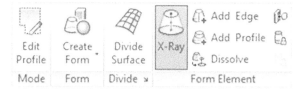

- You will see the following:

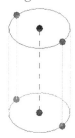

- Only one form can have X-Ray mode on at a time, and it displays in X-Ray mode in all views.
- Use commands like Edit Profile, Divide Surface, Add Edge, Add Profile, and Split Face to create more complex and accurate shapes:

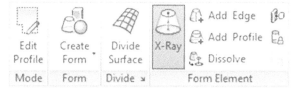

### Edit Profile

- You can edit the profile of the top of any form, as shown below. Start the command and select the desired profile; the selected profile will turn to a magenta color, and the rest will be grayed out. Make the necessary changes, then click (✓) to end the command:

### Add Edge and Profile

- You can add edges and profiles to forms. Start Add Edge command, hover your mouse over the desired side of form, and click to add the new edge. Same steps apply for adding new profile. You will receive the following:

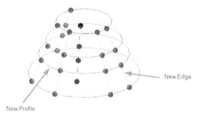

### Divide Surface

- You can divide any face of an In-Place mass to show a pattern as shown below. Also, you can split a face and assign materials for each part:

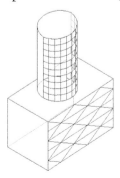

- How to Divide and assign a pattern to mass face:
  - While you are at In-Place editor, or in Conceptual Mass, select a face of In-Place mass.
  - In context tab, locate **Divide** panel, and click **Divide Surface** button:

  - The face will be divided into U and V distances.
  - Using the Type Selector, select a different pattern as shown here:

  - Using either Properties or Options Bar, edit the U and V distances, as shown here:

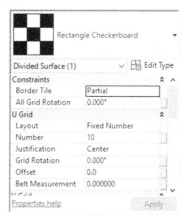

- How to Split mass faces:
  - Go to **Modify** tab, locate **Geometry** panel, and click **Split Face** button:

  - Select the desired face of the mass.
  - Using the **Modify** tab, locate the **Draw** panel, and use sketch tools to create the desired split (it should be closed inside the face, or open but with the condition of touching the edges of the face); when done, press [Esc] twice.
- How to apply material with the Paint command:
  - Using **Modify** tab, locate **Geometry** panel, and click **Paint** button:

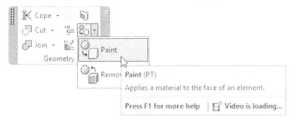

  - The Materials dialog box will display; select a material.
  - Hover your cursor over the desired face you want to paint, when it highlights it, click on the face to apply the material.
  - When you are done with assigning materials, click the **Done** button.
  - You can remove the material assignment, by going to the **Modify** tab, locate the **Geometry** panel, expand **Paint** and click the **Remove Paint** button, and select the face(s) you want to remove the material from:

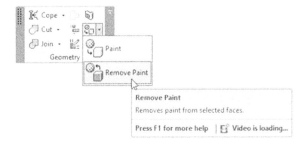

## EXERCISE 21-2 IN-PLACE MASSES

**1.** Start Revit 2023.

**2.** Open the file **Exercise 21-2.rvt**.

**3.** Switch on the **Show Mass Form and Floors** button.

**4.** Start the In-Place Mass command, and call the new mass "Offices Mass."

**5.** Make sure you are at the 00 Ground floor plan, and draw a rectangle 30000 (100'-0") width, and 25000 (84'-0") length.

**6.** Select the lines you draw, and Create Form. Go to the 3D view, set the height to be 8000 (20'-0"). Finish Mass creation.

**7.** Go to the 02 Second and Roof 1 floor plan view, set Underlay Range: Base level to 01 First.

**8.** Draw a reference plane from mid of the west edge, to mid of east edge. Draw another reference plane from north edge to south edge.

**9.** Start In-Place Mass command, and call the new mass "First Tower Mass."

**10.** Draw a circle (use Draw on Face option); the intersection of the two reference planes is its center, and Radius = 5000 (16'-8").

**11.** Go to 11 Roof 2 floor plan, and draw another circle with the same center, and Radius = 12000 (40'-0").

**12.** Go to the 3D view, select the two circles, and Create Form.

**13.** Finish Mass creation.

**14.** Go to the 02 Second and Roof 1 floor plan view. Draw more reference planes as shown here:

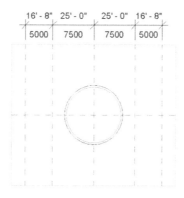

**15.** Start In-Place Mass command and call "Second Tower Mass."

**16.** Draw two arcs (using Start-End-Radius Arc command) and connect them using lines as shown in the following:

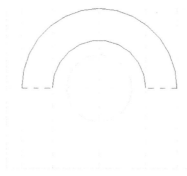

**17.** Select the two arcs and the two lines, and Create Form. Go to the 3D view.

**18.** Select the top face of the shape, then go to South elevation view, and drag it to the 07 Seventh level.

**19.** Go back to the 3D view.

**20.** Select the outer edge of the new shape, and set the radius to 15000 (50'-0").

**21.** Do the same thing for the inner edge, and set the radius to 10000 (33'-4").

**22.** Finish Mass creation.

**23.** You should receive the following shape:

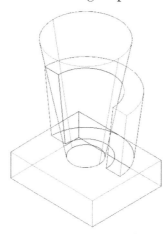

**24.** Go to the 00 Ground floor plan, and create a new reference plane, at the south edge of the existing mass, and name it **Entrance** just like the following:

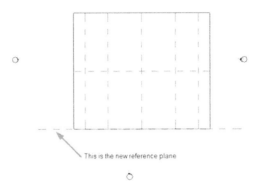

This is the new reference plane

**25.** Go to the South elevation.

**26.** Click the box, and from context tab click Edit In-Place.

**27.** Using the Work Plane panel, click Set button. When the dialog box appears from Name, select Entrance.

**28.** Draw a rectangle 25000 (83'-4") × 3500 (8'-0") using the leftmost and the rightmost reference planes as shown in the following:

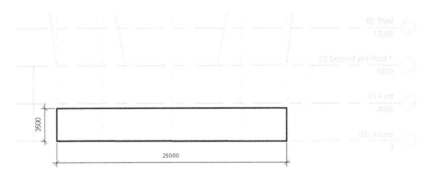

**29.** Go to the 3D view. Select the rectangle, and select Create Void; make sure that the void will be 5000 (16'-8") to the inside.

**30.** Finish Mass creation.

**31.** Save and close the file.

## CREATING WALLS, FLOORS, AND ROOFS FROM MASSES

- ■ The next step is to convert every part of the mass to walls (including curtain systems), floors, and roofs.
- ■ There are two ways for this conversion process to take place:
  - Go to **Massing & Site** tab, and locate **Model by Face** panel, as shown below:

  - For walls, using **Build** tab, locate Wall drop-down, and select **Wall By Face**. The same things apply to both Floors and Roof.
- ■ Either way, the conversion will take place.
- ■ The procedure is similar to all elements:
  - Open the suitable view which allows you to select the desired face.
  - Go to **Massing & Site** tab, locate **Model by Face** panel, and click Curtain System, Roof, Wall, or Floor.
  - Using the Type Selector, select the desired element type.
  - Select the desired faces, then select Create element type to finish (if you have more than one mass, then each element conversion will be in a separate command):

NOTE  *In order to create floors from your mass, you have to create* **Mass Floors** *as the first step.*

- ■ To create a Mass Floors, do the following steps:
  - Select a mass.
  - In context tab, locate **Model** panel and click **Mass Floors** button:

- You will see the following dialog box; select the desired levels in which you want to create floor area faces, as shown here:

- You will see the following:

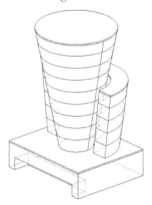

- If you select one of the Mass Floors, Properties will show the following:

- Mass Floors will calculate Floor Perimeter, Floor Area, Exterior Surface Area, and Floor Volume.

## EXERCISE 21-3 CREATING WALLS, FLOORS, AND ROOFS FROM MASSES

**1.** Start Revit 2023.

**2.** Open the file **Exercise 21-3.rvt**.

**3.** Select the three masses.

**4.** Using the context tab, click Mass Floors button.

**5.** Once the dialog appears, select all levels except 02 and 11.

**6.** Go to **Massing & Site** tab, locate **Model by Face** panel, and do the following:

    **a.** Use Curtain Systems for the two faces of the main tower.

    **b.** Use the CW 102-85-100p (Exterior - Brick on CMU) wall type for all the other walls.

    **c.** Use In situ Concrete 225mm (3" LW Concrete on 2" Metal Deck) for all floors.

    **d.** Use Warm Roof - Concrete (Generic - 12") for the three roofs.

**7.** Using the Visibility / Graphics dialog box, turn off Mass.

**8.** Visit different floor plans to look at your model.

**9.** Create two sections to cut your model horizontally and vertically to explore any mistakes in your design.

**10.** Save and close the project.

## NOTES

## CHAPTER REVIEW

1. Use _____ command to change the material of the mass if you do not like the transparent material assigned to it.

2. When you create a mass in Revit you will see it right away:

    **a.** True

    **b.** False

3. One of the following does not exist in Revit:

    **a.** Extrusion command

    **b.** Revolve command

    **c.** Loft Blend command

    **d.** Sweep command

4. Once you create a new mass, you cannot add more profiles and edges:

    **a.** True

    **b.** False

5. In order to create floors from your mass, you have to create _____ as the first step.

## CHAPTER REVIEW ANSWERS

1. Paint

3. c

5. Mass Floors

CHAPTER 22

# CUSTOMIZING WALLS, ROOFS, FLOORS, AND COMPOUND CEILINGS

**This Chapter Contains**

- How to create a new wall, roof, floor, and ceiling family
- How to create a vertically compound wall
- How to create a stacked wall
- How to create parts in walls

## INTRODUCTION

- This chapter will provide the knowledge for you to customize existing types of walls, floors, roofs, and compound ceilings.
- All of them are system families.
- The method is the same for all, except a wall has more options to include in the customization process.
- Walls will be used as an example for the four types.

## CUSTOMIZING BASIC WALL TYPES

- As previously stated, there are three types of walls; Basic, Curtain, and Stacked.
- We will cover the first and third in this chapter.

- In order to create a new basic wall, do the following steps:
  - Start the **Wall Architectural** command.
  - From Type selector, select a wall similar or close to what you want, and click **Edit Type** button. Using the dialog box, click **Duplicate** and type in a name for the new wall.
  - Under **Construction**, locate **Structure** parameter, and click **Edit** button. You will see the following dialog box:

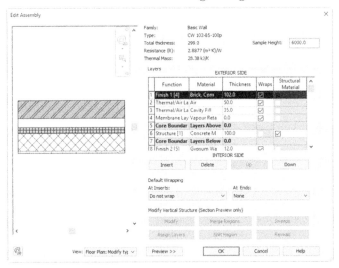

- **Assembly Information**: The top of the dialog box lists the Family, the Type, Total thickness, Resistance, and Thermal Mass. Also, you can see the Sample Height (this is important in case of a Stacked Wall).

- **Layers**: specify each layer of the wall, its function, thickness, whether to wrap or not, and whether it is Structural Material or not:

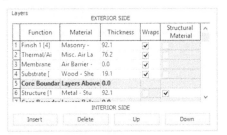

- Here is a list of the available functions that you can assign to wall layers:

| Structure [1] | This layer will be considered as support for other layers (it has the highest priority) |
|---|---|
| Substrate [2] | A material that acts as a foundation for another material, such as plywood or gypsum board |
| Thermal Air Layer [3] | A material which act as insulation and prevents air penetration |
| Membrane Layer | A membrane which will not allow water vapor penetration. This layer is set to a zero thickness, and has no priority code |
| Finish 1 [4] | The exterior finish layer |
| Finish 2 [5] | The interior finish layer (it has the lowest priority) |

**NOTE**

- *Priority codes will be used in wrapping. The priority 1 layer will wrap before the other higher priority code layers.*
- *There are four buttons beneath the table, to Insert, Delete, move Up, and move Down layers.*
- *The top of the table is labeled Exterior, and the bottom is labeled Interior.*
- *You may have more than one Core Boundary layer in your wall.*

### Wrapping (Only for Walls)

- **Default Wrapping**: This is None At Ends (end of walls from both sides), and Do not wrap At Inserts (when door or window is inserted):

- **For At Ends**: The available options are: None, Exterior, or Interior:
  - Exterior means all layers above Core Boundary will be wrapped at the two ends:

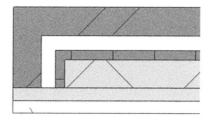

- Interior means all layers below Core Boundary will be wrapped at the two ends:

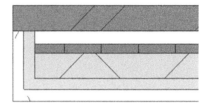

- **For At Inserts**: The available options are: Do not wrap, Interior, Exterior, or Both:
  - Interior means all layers below Core Boundary will be wrapped at the insert of the door or window:

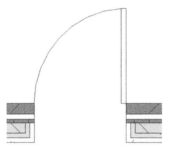

  - Exterior means all layers above Core Boundary will be wrapped at the insert of door or window:

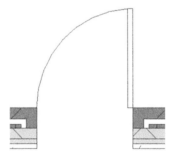

- Both means all layers below and above Core Boundary will be wrapped at the insert of door or window:

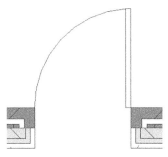

## USING PROFILES ON WALLS, ROOFS, AND FLOORS

- Revit has many commands which enable you to add profiles to walls (Sweeps and Reveals), roofs (Soffit, Fascia, and Gutter), and floors (Slab Edges).
- Profile is a 2D cross section which can be applied to the definition of walls, roofs, and floors.
- We can use premade profiles or you can customize your own.
- Go to **Architecture** tab, locate **Build** panel, as shown below, and select the desired type you want to add the profile to:

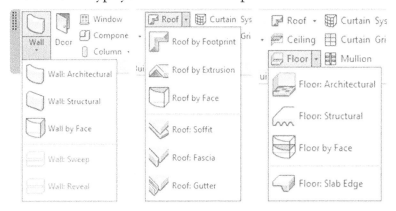

- To set up profiles, and add it to the element, do the following steps:
  - Go to elevation or 3D view.
  - Go to **Insert** tab, locate **Load from Library** panel, and click **Load Family** button.

- Locate the Profiles folder, choose the desired folder (Walls, Roofs, Slabs, etc.), select the profile, and load it.
■ To apply the profile, and as an example, we will discuss Sweep, but the procedure fits for other types as well:
  - Start **Sweep** command. Using Type selector, select a Wall Sweep type and click **Edit Type** button. In the Type Properties dialog box, click the **Duplicate** button and enter a new name for the type.
  - In the Type Properties dialog box, under **Construction**, select **Profile** (which you already load it to the project), as shown below:

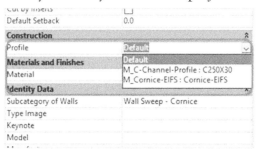

- Set the Material of the Sweep. When done, click OK.
- Using context tab, locate **Placement** panel, click either **Horizontal** or **Vertical**. (This is only for walls.):

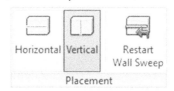

- Hover cursor over an element to add the sweep and click to place it.
- When done press [Esc] twice.
- You can adjust the height (for vertical) and width (for horizontal) by clicking them and moving the blue circle.

## EXERCISE 22-1   CREATING BASIC WALL TYPES

**1.** Start Revit 2023.

**2.** Open the file **Exercise 22-1.rvt**.

**3.** Start the Wall command, and make sure that Generic – 200mm (Generic 8") is selected, duplicate it, and name the new wall type **My Wall**.

**4.** Add and delete layers to create the below wall type:

**5.** Set the Location Line to be Finish Face: Interior.

**6.** Using the existing rectangle, draw a rectangle wall. Go to the 3D view to make sure that the Brick layer is at the outside.

**7.** Go to back to the Level 1 view. Make sure you can see all layers of the wall.

**8.** Add a door at the north wall. See how the layers wrap (no wraps by default).

**9.** Select the wall and click the Edit Type button.

**10.** Under Construction, locate Wrapping at Inserts, and change it to Exterior.

**11.** Click OK and see how the exterior layers only wrapped at inserts.

**12.** Start the wall command again, using My Wall type, draw a 5000mm (20'-0") wall above the north wall. Check how the two ends of the wall are not wrapped.

**13.** Select the wall and click the Edit Type button.

**14.** Under Construction, locate Wrapping at Ends, and change it to Exterior.

**15.** Click OK, and see how the exterior layers only wrapped at the ends.

**16.** Go to **Insert** tab, locate **Load from Library** panel, and click **Load Family** button.

**17.** Go to Profiles / Wall, and load M_Cornice-Precast.rfa (Cornice-Precast.rfa).

**18.** Go to the 3D view.

**19.** Start the **Wall Sweep** command.

**20.** Click Edit Type, click Duplicate, and name the new sweep My Sweep.

**21.** Under Construction change the profile to be M_Cornice-Precast (Cornice-Precast.rfa).

**22.** Add an outside sweep to all walls except the north wall, making sure that the sweep is 1000 mm (3'-4") from top.

**23.** You should receive the following:

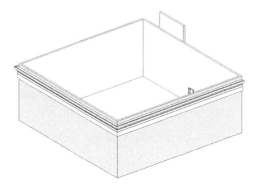

**24.** Save and close the project.

## VERTICALLY COMPOUND WALLS

- You can create a wall of multiple materials in a single layer.
- Also, you can add permanent sweeps or reveals.
- Inside the Edit Assembly dialog box, check the following:
  - Set Sample Height value
  - Change the view to Section
- Refer to the following illustration:

- Do the following steps:
  - Because View is Section in the Preview part, the Modify Vertical Structure tools will be activated as shown below:

  - Use zooming in the section part as desired to see the various layers.
  - Click the Split Region button and the mouse pointer will change to:

- In the section part, move closer to the wall, until you see a dimension. Move up and down to locate the exact distance of the intended split, then click. You will see small horizontal line indicting a split. In the table, the thickness of that layer will change to Variable.

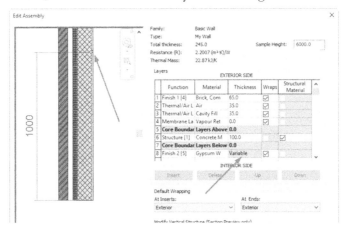

- Create a new layer and assign material to it without assigning a thickness.
- Select the new material (using the number at the leftmost), click the Assign Layers button, and click the desired split.
- The two layers now will have the same thickness.

| | Function | Material | Thickness | Wraps | Structural Material |
|---|---|---|---|---|---|
| 2 | Thermal/Air L | Air | 35.0 | ☑ | |
| 3 | Thermal/Air L | Cavity Fill | 35.0 | ☑ | |
| 4 | Membrane La | Vapour Ret | 0.0 | ☑ | |
| 5 | **Core Boundar Layers Above** | | **0.0** | | |
| 6 | Structure [1] | Concrete M | 100.0 | | ☑ |
| 7 | **Core Boundar Layers Below** | | **0.0** | | |
| 8 | Finish 2 [5] | Cherry | 10.0 | ☑ | |
| 9 | Finish 2 [5] | Gypsum W | 10.0 | ☑ | |

- If you want to change the height of the split, simply zoom in until you can see the small line representing the split; once you select it, the dimension will turn to temporary dimension, input the new value.
- To merge layers together, do the following steps:
  - Click the **Merge Regions** button.

- Select the line between the two desired layers you want to merge, as shown below:

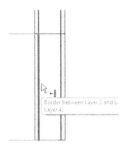

- Check the arrow cursor, it will indicate to you which way the merge will occur. Also, check the tool tip which will list the names of the layers that are merging.
- You can only merge layers that are next to each other.
- You can merge layers vertically or horizontally.

### Permanent Wall Sweeps and Reveals

- To add a permenant sweep or reveal like shown below, try the following:

- Make sure that your desired profile is already loaded.
- Click the **Sweeps** button, and a dialog box, as shown below, will open:

- Click the **Add** button to add a profile and assign material. Then set the Distance from the base or top, interior or exterior Sides, and the Offset from that side. Click Flip checkbox on, to flip the profile upside down.

## EXERCISE 22-2    CREATING VERTICALLY COMPOUND WALLS

1. Start Revit 2023.

2. Open the file **Exercise 22-2.rvt**.

3. Go to the 3D view, and select one of the walls. Click Edit Type, locate Structure, and click Edit.

4. In Edit Assembly dialog box, change the Sample Height to 3600mm (12'-0").

5. Click **Preview** button and set the view to the **Section: Modify type attributes** option.

6. Add a new layer beneath layer # 8, set it to be Finish 2, and assign material **Cherry** to it. Don't set the thickness.

7. Click the Split Region button, and split the Gypsum Wall Board layer at height = 1200mm (4'-0") (zoom in, and make sure here is a small horizontal line in the thin layer). Check if the thickness of this layer became **Variable**.

8. Select row # 9 (the new material). Click the Assign Layers button, and select the lower split.

9. Click the Modify button to make sure that the lower part became a different material. Also, check if the two material has a read-only thickness value.

10. Click the **Sweeps** button.

11. Click the **Add** button and set up one wall sweep using the following information:

    a. Profile: M_Cornice-Metal Panel : 500 × 750mm (Cornice-Metal Panel : 20" × 30")

    b. Distance: 1950mm (6'-6")

    c. From: Base

    d. Side: Interior

**12.** Click OK twice. Set the visual style to Shaded and look at the model from different angles.

**13.** Go to Level 1, and create a wall section in one of the walls, and understand it.

**14.** Save and close the project.

## VERTICALLY STACKED WALLS

- Stacked wall is the third type of wall, after Basic and Curtain wall types.
- You can stack two or more basic walls to get a stacked wall type.
- In default template, you will find one stacked wall type which can be duplicated and modified.
- One of the basic wall types in the stacked wall should have Height=Variable, the other types can have fixed height.
- To create a Vertically Stacked Wall, do the following steps:
  - Start the Wall command.
  - Select the existing stacked wall type, click the **Edit Type** button, click the **Duplicate** button, and give the new type a new name.
  - Click the **Edit** button next to the Structure parameter. In the Edit Assembly dialog box, set the Offset for how the walls are stacked, then set the Sample Height of the wall, as shown below:

  - Using the Types table, choose which basic wall types you want to add to the stacked wall, as shown below:

- For each wall type, set the desired height. One height must be variable. Set the Offset of the wall as needed.
- Click the **Preview** button to see the wall.
- Click OK until all dialog boxes are closed.

### Embedding a Wall Inside Another Wall

- The only wall which can be embedded in other wall types is Storefront.
- But there is a small trick which will allow any basic wall to be embedded in any other basic wall.
- Do the following steps:
  - Create a basic wall from any type, we will call it the host wall.
  - Using the Base offset and Tope offset, add the embedded wall to the host wall.
  - You will see a warning appear:

  - Close the warning box.
  - Go to the **Modify** tab, locate the **Geometry** panel, click the **Cut** drop down, and select the **Cut geometry** command. Select the host wall first, then select the embedded wall.
  - Look at the wall in 3D, elevation, or section views; using controls you can modify the height of the embedded wall.

# EXERCISE 22-3    CREATING STACKED AND EMBEDDED WALLS

1. Start Revit 2023.

2. Open file **Exercise 22-3.rvt**.

3. Start the Wall command.

4. From the Type selector, select the only available stacked wall type. Click the Edit Type button, Duplicate, and name it My Stacked Wall.

5. Next to Structure, click the Edit button.

6. Change offset to Wall Centerline, and Sample Height = 8000mm (20'-0").

**7.** Create a stacked wall as shown in the following:

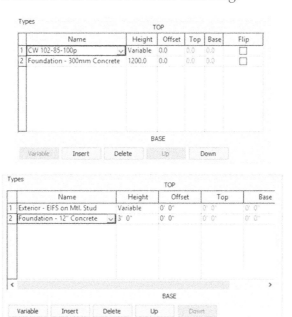

**8.** Click Preview to see the stacked wall. Click OK twice to end the command.

**9.** Set the Base level = Level 1, and Top level = Level 3, set the Location line to be Finish Face: Interior.

**10.** Using the existing rectangle, draw a rectangle wall. Go to the 3D view, and look at the model.

**11.** Go back to the Level 2 floor plan view.

**12.** Start the wall command again, and select Generic – 200mm (Generic – 8").

**13.** Set the Base level = Level 2, and Top level = Level 3, and set the Location line to be Wall Centerline.

**14.** Draw a wall on the centerline of the north wall; the distance is not important.

**15.** Read the warning, and close it. Start the Cut Geometry command. Select the north wall first, then select the new wall. Go to 3D to see the result.

**16.** Save and close the file.

## CREATE PARTS

- You can cut a single wall (floor, roof, and compound ceiling) layer to parts, using any drawing tool.
- You can ask Revit to remove any part of this layer.
- Create Parts will help you create a view showing all the layers inside a wall, floor, roof, or compound ceiling.
- You cannot create parts on stacked walls or curtain walls.
- To create parts, do the following steps:
  - Select the desired wall, floor, roof, or compound ceiling.
  - Go to **Modify** tab, locate **Create** panel, and click **Create Parts** button:

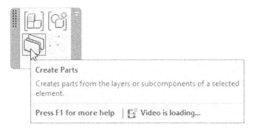

  - You can now see each layer by itself.
  - The context tab will look similar to the following:

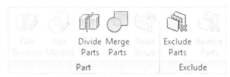

  - Click the **Divide Parts** button to cut the desired element either vertically, or horizontally. The context tab will be as follows:

  - Click the **Edit Sketch** button, so you can draw a 2D shape for Revit to cut.
  - If you are dealing with a wall, the plane will be parallel to the plan view. If you want to work on the face, locate the **Work Plane** panel, click the **Set** button, select a **Pick a plane** choice, and click the face

of the wall. The working plane is right in the case of floor, roof, or compound ceiling.

- Once you are done click (✔).
- Look at Properties and you will see the following:

- If you want to set a gap between the parts, input a value.
- If you want to have a division profile, set it from the pre-loaded profile (you can load a profile beforehand).
- Set the different requirements of the profile:

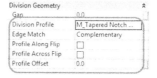

- Once you are done click (✔).
- You will receive the following:

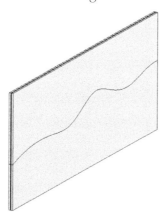

- If you click on one of the parts, Properties will show everything related to this part, as shown below:

- You can see under Dimensions information like: Volume, Area, Length, Height, Thickness, whether to be excluded or included.
- Under Identify Data, you can see whether or not to Show Shape Handle. What is the original Category, Family, and Type. Whether to keep the original material or to assign a new material for it.
- If you click on one of the parts, the context tab changes to:

- Using the Edit Division button, you can edit the Division again.
- Using Divide Parts, you can divide the part further.
- Using Exclude Parts, you can exclude the selected part (it will disappear). Restore Parts will retrieve it.
- If you select two parts or more, Merge Parts will be enabled, so you can merge the two or more selected parts. Hence, the Edit Merged button will be enabled.
- Reset Shape will be enabled only after you show shape handles (from Properties) and change the part. It will reset it back as if no change was made.

# EXERCISE 22-4   CREATE PARTS

1. Start Revit 2023.

2. Open file **Exercise 22-4.rvt**.

3. Select the wall.

4. Start **Create Parts** command. Now you can see the layers of that wall. Press [Esc] twice to end the command.

5. Select the brick layer. Click Divide Parts command, then click Edit Sketch button.

6. Using the Work Plane panel, click the Set button. Select Pick a plane. Click OK, and select the outer face of the brick layer.

7. Start the Spline drawing tool. Almost at the middle of the height of wall draw a series of splines to cut it into two halves. When done, click (✓).

8. From Properties, set the Gap = 500 (0'-20"), click (✓).

9. You will receive the following:

10. Select the lower part and click the Exclude Parts button.

11. Select the upper part and check Properties for Area and Volume.

12. Reselect the lower part and click Restore Parts.

13. Zoom out and select the other wall at the right.

14. Select the wall, click Create Parts button. Now you can see all layers. Press [Esc] twice to end the command.

**15.** Select all layers (using crossing).

**16.** From Properties turn on Show Shape Handles.

**17.** Select the outer layer, and stretch it to the inside to create a smaller representation of this layer.

**18.** Do the same for the other layers. You will receive the following:

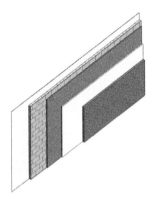

**19.** Save and close the file.

## NOTES

## CHAPTER REVIEW

1. In stacked wall, one height must be _____.

2. Create Parts will help you create a view showing all the layers inside a floor:

   **a.** True

   **b.** False

3. Membrane layer in wall definition:

   **a.** Will not allow water vapor penetration

   **b.** Is set to a zero thickness

   **c.** Has no priority code, you can select one element of the group and hide it

   **d.** All of the above

4. Profile is a 2D cross section which can be applied to the definition of walls, roofs, and floors:

   **a.** True

   **b.** False

5. While you are in the wall definition dialog box you can:

   **a.** Split Region

   **b.** Add a permanent sweep and reveal

   **c.** Roof Fascia and Gutter

   **d.** Set up wraps for Insert

6. To allow any basic wall to be embedded in any other basic wall, select _____ command.

## CHAPTER REVIEW ANSWERS

1. Variable

3. d

5. c

## CHAPTER 23

# CREATING FAMILIES IN REVIT

**This Chapter Contains**

- What is the best way to create a family in Revit
- Creating a framework of family
- Creating a 3D body of the family
- Creating types and assigning materials

## INTRODUCTION TO CREATING FAMILIES

- Revit is equipped with many component families from furniture, lighting, doors, windows, etc.
- We practiced how to create System families like Walls, Floors, Roofs, and Compound Ceilings.
- You can create your own component family by using Family Editor, and 3D creation commands like the ones we saw in Mass Modeling.
- Families can be:
  - **Host-based families** like light components, they need walls or a ceiling to host them
  - **Stand-alone families** like furniture components, they do not need a host

- To create a new component family, do one of the following two:
  - Start from an empty template file (for each type there is a separate template file)
  - Open an existing RFA, save it under another name, and make the necessary changes
- Either way, you will use the Family Editor as your interface. Family Editor is equipped with all tools to accomplish your mission.

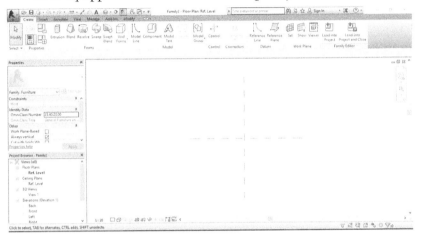

- Before you start your first step, try the following:
  - Create sketches of your family from top and side views, mentioning all the needed dimensions.
  - Specify which dimensions will be variable, and which will be constant.
  - Since our family is parametric, then you need to draw imaginary reference lines which will define the framework of your family.

## PARAMETRIC FRAMEWORK

- Revit families are parametric.
- As a first phase to create a family, do the following steps:
  - Add to the existing reference planes (they include the template file) the desired new reference planes
  - Put dimensions (use EQ to specify equal distance)
  - Create a parameter by naming the dimension with relative names
  - Create relationship between parameters
  - Flex your framework, by inputting different values for certain parameters, and see how the other parameters reacted to the change

- Check the shape below:

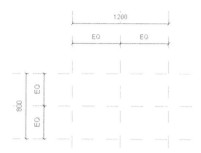

- To draw Reference Planes, do the following steps:
  - In the Family Editor, using **Create** tab, locate **Datum** panel, and then click **Reference Plane** button:

  - The method of creating a reference plane was already covered in the beginning of this book.
- Reference Lines are continuous green lines, which define four work-planes, as follows:
  - Plane parallel to the workplane of the line itself
  - Plane perpendicular to the first
  - Two planes at the two ends of the line
- You can see reference planes/lines in plan and elevation views but not in 3D views.
- Once you start a new family, you will see two intersecting reference planes. This will be considered the origin of the family.
- If you draw another reference plane and you want to change the origin of the family, simply select the desired reference planes, using Properties, locate Other, and then click the checkbox **Define Origin** on:

- This will impact the Room Calculation Point (by default hidden). The Room Calculation Point is the point that Revit uses in order to create the furniture schedule per room. The Room Calculation Point will always be following the Origin of the family.

■ To show the Room Calculation Point, using Properties, under Other, click the checkbox on as shown below:

■ *As discussed previously, you can give reference planes names.*
■ *Reference Planes are not constraints unless you put dimensions on them.*
■ *In order to create relationships between these dimensions, you have to convert them to parameters first.*

■ To label a dimension, do the following steps:
  • Select the dimension you want to label
  • Using context tab, locate **Label Dimension** panel, and click **Create Parameter** button to create a new label:

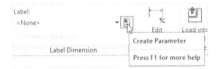

  • If the parameter is already created, you will find it in the list, simply select it as shown below:

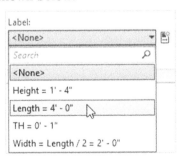

- If you are adding a new parameter, the Parameter Properties dialog box will be shown. Select the Family parameter option, as shown here:

- In the Parameter Data section, type in the Name. The Type of the Parameter is by default set to Length and the Group parameter under is set to Dimensions, because the parameter is a dimension:

- Select Type or Instance, then click OK.

**NOTE**

- *Revit will issue a warning if there are too many parameters. The reason is, it will be over-constraining the family. Hence, the new parameter will not be created.*
- Test your framework, by inputting different values for one of the parameters, and see how the other parameters react. This is called falexing your parameters.
  - Go to **Create** tab, locate **Properties** panel, and click **Family Types** button:

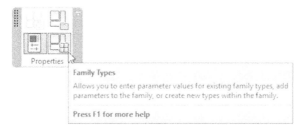

- You will see the following dialog box:

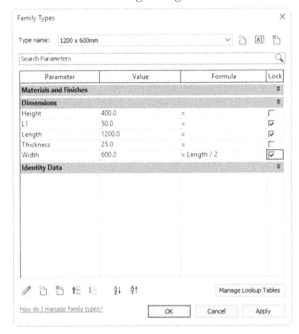

- Use the Value field to change the numbers.
- Add formulas under the Formula column. The example above is Width = Length/2.
- Use the following in formulas:
  - Addition (+), subtraction (-), multiplication (*), division (/), exponentiation, logarithms, and square roots.
  - Sine, cosine, tangent, arcsine, arccosine, and arctangent.
  - Less than (<), greater than (>), equal (=). Also, use AND, OR, or NOT.

**NOTE** *Parameters are case-sensetive. If you named one of your paramater Widths (w is capital) you have to input it like that in the formula.*

## EXERCISE 23-1    COFFEE TABLE FRAMEWORK

**1.** Start Revit 2023.

**2.** Using the File menu, click **New**, then select **Family** option.

**3.** Select the template Metric Furniture.rft (Furniture.rft) and click Open button.

**4.** At the Project Browser, you are at Floor Plans, Ref. Level. You can see two existing reference planes are included in the file.

**5.** Save the family in your Exercise folder as Coffee_Table.rfa.

**6.** Add two new vertical reference planes, one at the right, and the other at the left of the existing plane. Do the same for the horizontal:

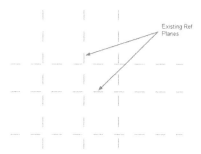

**7.** Put dimension between the three vertical reference planes, setting EQ. Then add single dimension from the first to the third. Do the same for the horizontals.

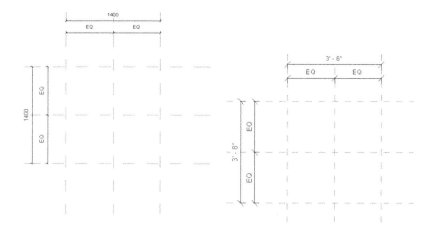

**8.** Create from the horizontal dimension a parameter label named **Length**. Create from the vertical dimension a parameter label named **Width**. You should receive the following:

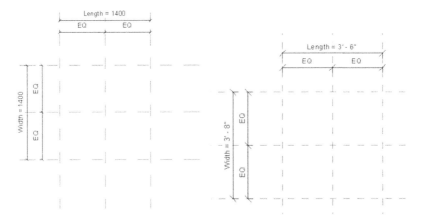

**9.** Click Family Types button and do as shown below:

| Parameter | Value | Formula | Lock |
|---|---|---|---|
| **Dimensions** | | | ☆ |
| Length | 1200.0 | = | ☑ |
| Width | 600.0 | = Length / 2 | ☑ |
| Identity Data | | | ✕ |

| Parameter | Value | Formula | Lock |
|---|---|---|---|
| **Dimensions** | | | ☆ |
| Length | 4' 0" | = | ☑ |
| Width | 2' 0" | = Length / 2 | ☑ |

**10.** Go to Front view. Create two reference planes and then dimension and label them as shown below:

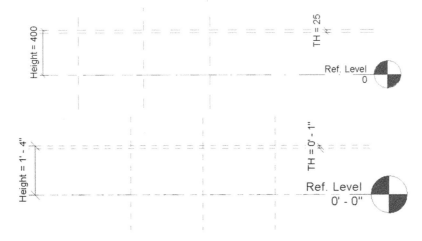

**11.** Using Family Types flex the framework, by inputting values in the length, 1500 (5'-0"), and 1800 (6'-0"). Get it back to 1200 (4'-0")

**12.** Save the family file.

## CREATING THE 3D ELEMENTS OF THE FAMILY

- Your next step is to create the 3D elements
- There are five commands to create 3D solid elements, as shown here:

- Each method uses a sketched 2D profile as the basis of the 3D shape:
  - **Extrusion command** creates a 3D solid by extruding a 2D profile
  - **Blend command** creates a 3D solid by blending two 2D profiles
  - **Revolve command** create a 3D solid by rotating a 2D profile around an axis
  - **Sweep command** creates a 3D solid by sweeping a 2D profile along a path
  - **Swept Blend command** creates a 3D by sweeping two 2D profiles along a path
- Use **Align** command to align and lock the 3D solid elements to the existing reference planes.

- To create a constant distance, use dimension commands, and lock them.

**Adding Controls**

■ Adding Controls is very handy in a doors and windows family, where you can assign them to flip the door right to left, and inside out, as shown here:

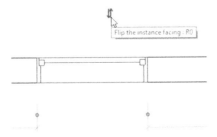

■ To add controls, do the following steps:
  • Go to **Create** tab, locate **Control** panel, and click the **Control** button:

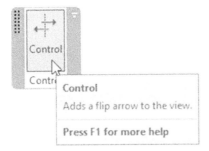

  • In the context tab, locate **Control Type** panel, and click one of the following buttons:

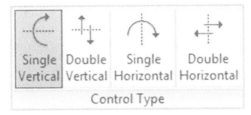

  • Click on the screen to place the control in the view.
  • Press [Esc] twice to end the command.

### Adding Openings

- In case you started a family which needs a host (like doors and windows) opening command will allow you to draw an opening in a host element.
- To add an opening to a host element, do the following steps:
  - Go to **Create** tab, locate **Model** panel, and click **Opening** button:

  - In the context tab, locate the **Draw** panel, and using the Draw tools sketch the opening. When done, click (✓).
  - In the context tab, make changes as needed. Using the Options Bar, select whether you want the cut to display in 3D and/or in Elevation.
  - Press [Esc] twice to finish the command.

### Adding Model and Symbolic Lines

- To draw 2D elements to help see some more details in certain views, use one of the following two commands:
  - Go to **Create** tab, locate **Model** panel, then click **Model Line** button:

  - Go to **Annotate** tab, locate **Detail** panel, and then click **Symbolic Line** button:

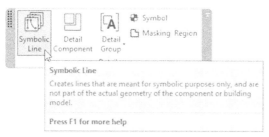

- You can see Model Lines in 2D and 3D views.
- However, you can add Symbolic Lines in views parallel to the view you created them in (a good example would be to use Symbolic lines to describe the swing of a door).

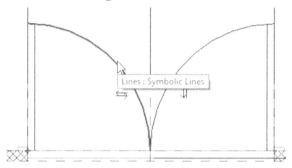

**NOTE** *You can add Text and Model Text to your family as you did with a normal project.*

## EXERCISE 23-2    CREATING 3D ELEMENTS OF THE FAMILY

**1.** Start Revit 2023.

**2.** Open the file **Exercise 23-2.rfa**.

**3.** Go to Ref. Level floor plan view.

**4.** Start Extrusion command.

**5.** Set the Depth = 25mm (0'-1").

**6.** Start the Rectangle from the Draw panel, and draw a rectangle 1200 × 600 mm (4'-0" × 2'-0"). Lock the four lines to the reference planes. Click (✓).

**7.** Go to Front elevation view. You will notice that the top of the coffee table is not at the right place. Using Align command to align the top edge to the top reference plane, and lock it. Then align the bottom edge to the lower plane, and lock it as well.

**8.** Go to 3D view, and you will see the top of the coffee table.

**9.** Go to Ref. Level floor plan view.

**10.** To place the legs of the coffee table, we need to create four reference planes, which they are an offset of 50mm (0'-2") to the inside of the four edges.

**11.** Put dimensions and *lock* them for the four new reference planes. You should receive the following:

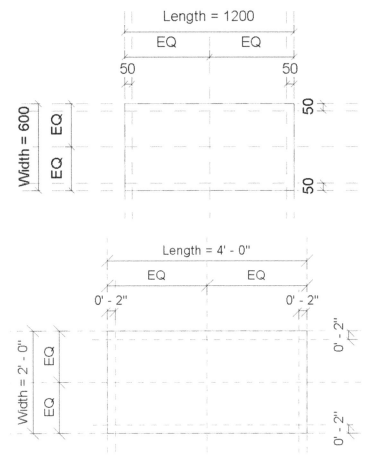

**12.** Create another four reference planes with offset of 50mm (0'-2") to the inside measured from the new four reference planes.

**13.** Put dimensions and *lock* them for the four new reference planes. You should receive following:

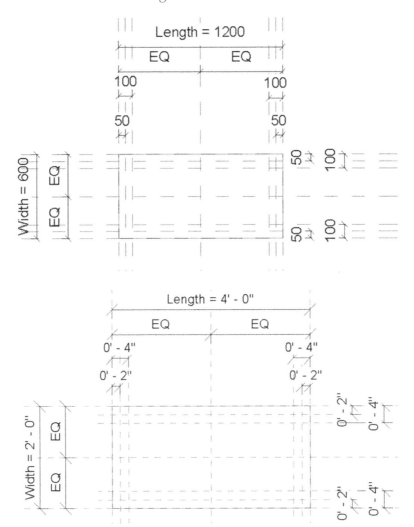

**14.** Zoom to the upper right corner of the tabletop.

**15.** Start Blend command, pick the Circle from Draw panel, and draw R=20mm (0'-0 ¾") circle in the intersection. Set the Depth to 375mm (1'-3"). From the context tab, locate the Mode panel, and click the Edit

Top button. The Circle is still with you; draw another R=40mm (0'-1 ½")
circle in the other intersection. This should be the case:

**16.** Click (✔).

**17.** Look at your family using the 3D view. You should have one leg.

**18.** Go back to Ref. Level floor plan view.

**19.** Using the Mirror – Pick Axis command, mirror the leg twice to get four
legs.

**20.** Go to the 3D view, you should see the following:

**21.** Save and close the file.

## FAMILY TYPES – SIZES AND MATERIALS

- As a final step, we will create preset sizes to be used by the user.
- Then we will create a material parameter which will give us the flexibility
  to assign materials to different parts of our family.

**Setting Sizes**

- To create family sizes, do the following steps:
- Go to **Create** tab, locate **Properties** panel, and select **Family Types** button:

- You will see the following dialog box:

- Click **New Type** button:

- The following dialog box will appear; type in the name of the new type, and click OK:

- Input the values for that size, as shown below:

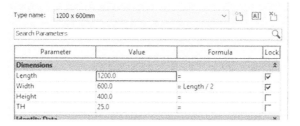

- Click Apply to check the size without leaving the dialog box.
- Do the same for other sizes; when done, click OK to end the command.
- Save and load the family to your project.
- Use the two buttons at the right of New button, to **Rename** and **Delete** the existing types.

### Assigning Materials

- There are two ways to assign material for all or part of your family:
  - Select the part, using Properties, locate Material parameter, and assign a material for this part. Using this method, the part of your family will always carry the selected material.
  - The other method will give you the flexibility to let the user of your family assign material as type or instance parameters.
- The following steps will let you assign materials as parameter:
  - Select the desired part of your family.
  - Using Properties, locate Materials and Finishes, then locate Material. You will see >By Category<. Click the small button at the right as shown below:

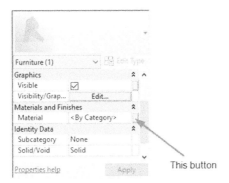

- You will see the following dialog box:

- As you can see no material parameters were defined before. Click the **Add** parameter button. You will see the following dialog box:

- Type in the Name. Select whether it will be a Type or Instance parameter. The Type of Parameter is preset to be Material, and Group parameter under is set to Materials and Finishes.
- When done click OK.
- The previous dialog box will have this parameter added; select it and click OK.
- Go to the Family Types dialog box.
- Under Materials and Finishes, you will see the following:

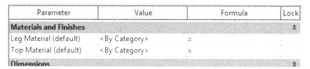

| Parameter | Value | Formula | Lock |
|---|---|---|---|
| **Materials and Finishes** | | | ☆ |
| Leg Material (default) | <By Category> | = | |
| Top Material (default) | <By Category> | = | |
| **Dimensions** | | | ☆ |

- Assign materials for all the material parameters as desired.
- When done click OK.

## EXERCISE 23-3  FAMILY TYPES – SIZES AND MATERIALS

1. Start Revit 2023.

2. Open the file **Exercise 23-3.rfa**.

**3.** Click Family Types, and add three new types:

    **a.** 1200 × 600mm (4' × 2') (the default dimensions)

    **b.** 1500 × 750mm (5' × 2'-6") (input Length = 1500, 5')

    **c.** 1800 × 900mm (6' × 3') (input Length = 1800, 6')

**4.** Make sure that the 1200 × 600mm (4' × 2') type is selected and click OK.

**5.** Select the top of the table.

**6.** Using Properties, locate Materials and Finishes, locate Material, and click the small button to the right. Add a new parameter:

    **a.** Family parameter

    **b.** Name = Top Material

    **c.** Instance parameter

    **d.** Click OK twice

**7.** Select the four legs.

**8.** Using Properties, locate Materials and Finishes, locate Material, and click the small button to the right. Add a new parameter:

    **a.** Family parameter

    **b.** Name = Legs Material

    **c.** Instance parameter

    **d.** Click OK, when done

**9.** Go to Family Types dialog box, and assign Top Material = Glass, and Legs Material = Aluminum.

**10.** Save the family as myname_coffee_table.rfa.

**11.** Create a new file and load the family to it. Try the different sizes and different materials.

# NOTES

## CHAPTER REVIEW

**1.** In Family Editor, when you create a dimension, you create a parameter:

    **a.** True

    **b.** False

**2.** Use _____ command to align and lock the 3D solid elements to the existing reference planes.

**3.** Type parameter is the same as Instance Parameters:

    **a.** False

    **b.** False, not always, in some cases

    **c.** True

    **d.** True in some cases, and false in other cases

**4.** Loft is not among the commands you can use in Family Editor:

    **a.** True

    **b.** False

**5.** While creating a family in Revit, one of the statements are not true:

    **a.** Assign material for each part of the family

    **b.** Let the user decide which material is the right material for each part

    **c.** Create preset sizes

    **d.** You cannot define formulas between parameters

**6.** _____ families like furniture components, they do not need a host.

## CHAPTER REVIEW ANSWERS

    **1.** b

    **3.** a

    **5.** d

## CHAPTER 24

# CUSTOMIZING DOORS, WINDOWS, AND RAILING FAMILIES

## This Chapter Contains

- Customize Door family
- Customize Window family
- Customize Railing system family

## CUSTOMIZING EXISTING DOORS AND WINDOWS

- Revit includes many doors and windows families.
- These families help you fulfill the requirements of almost all of your projects. Still, we need to know how we can customize the existing doors or even create a door from scratch.
- You have two ways to go:
  - Open an existing door / window family, and save it under a new name. Make all the changes needed by adding reference planes, adding new 2D/3D elements, and removing any unnecessary elements. This is the recommended way.
  - Start a new family based on one of the door RFT templates that is included with Revit. This method is harder and longer than the previous one.
- We will do two exercises which will use the first method, one for doors and the second for windows.

## VISIBILITY SETTINGS

- Visibility Settings allow the user to control when elements of your family will be shown, and when they will be hidden.
- Certain objects like handles of doors should not be displayed in floor plans or ceiling plans.
- To control the display of elements, do the following steps:
  - Select the desired element(s). Using context tab, locate **Mode** panel and click **Visibility Settings** button:

  - You will see the following dialog box; specify the views and detail level in which you want the elements to display, as shown below:

- The dialog box states that all elements will be displayed in 3D. Select other view(s) in which you want the elements to be shown.
- Also, select which detail level you want the element to be shown in. For example, maybe you want to show an element in floor plans but not in Coarse detail.

## EXERCISE 24-1 CREATING A CUSTOM DOOR

**1.** Start Revit 2023.

**2.** From the starting page, under Families, click Open.

**3.** Go to Doors folder and select M_Door-Single-Panel.rfa (Door-Single-Panel.rfa), and click Open.

**4.** Save the file (in your exercise folder) as My_Door.rfa.

**5.** Become acquainted with the family by visiting Ref. Level in Floor Plans, 3D view, and the four-elevation view. Also, visit the Family Type dialog box to understand the parameter names, and what each one represents.

**6.** Go to Front elevation view.

**7.** Create two new vertical reference planes by offsetting the existing middle reference planes using 250mm (0'-10"), one to the right, and one to the left.

**8.** Create three new horizontal reference planes by offsetting the upper horizontal reference plane by 500mm (0'-20") three times.

**9.** Input the dimensions and names as shown below:

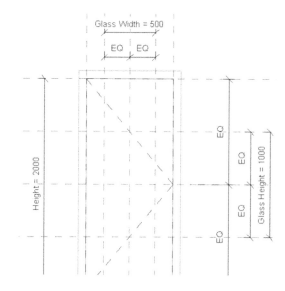

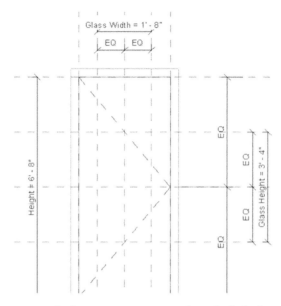

**10.** Go to **Create** tab, locate **Forms** panel, and click drop-down list of Void Forms; select Void Extrusion.

**11.** Set Depth value to 250 (0'-10"), draw a rectangle as shown below and lock the four edges:

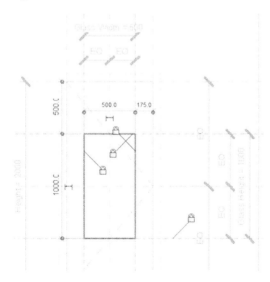

**12.** Click (✔) to end the command.

**13.** Go to 3D view.

**14.** Select the void, and extend it to penetrate the door fully.

**15.** Go to the Cut Geometry command, and select the door, then the void. Press [Esc] twice.

**16.** This is what you should receive:

**17.** Go to Ref. Level view.

**18.** Create at the middle of the door (not the wall) a new reference plane, and call it Center of Door.

**19.** Create two more reference planes using an offset value of 6mm (0'-0 ¼"), one above it and one below it.

**20.** Input the dimensions and names as shown below:

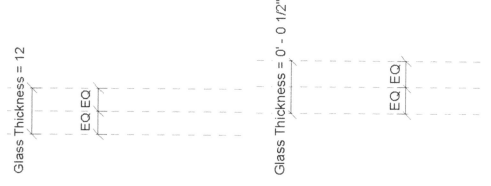

**21.** Start the Extrusion command, and draw a 12mm (0'-0 ½") thickness rectangle that covers the opening in the door, and lock the four sides. Set the

Extrusion Start = 500 (1'-8"), and Extrusion End = 1500 (5'-0"). When done click (✓).

**22.** Select the newly created extrusion and assign Glass as the material.

**23.** Go to the 3D view to check your model.

**24.** Go to Ref. Level view.

**25.** Go to the Create tab, locate the Model panel, and click Component button. Click Load Family, select Doors, then Hardware, and select M_Handle Domestic.rfa (Handle Domestic.rfa), and click Open.

**26.** Press Spacebar twice to rotate it.

**27.** Using Properties, set Offset to 1000mm (3'-4"), and insert it, as shown in the following:

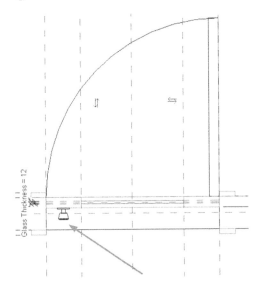

**28.** Select the extrusion, using the context tab, locate the Mode panel Visibility Settings button, and turn off Plan/RCP check box, and When cut in Plan/RCP and click OK.

**29.** Repeat the same for the handle.

**30.** Save and load the door into a project for testing.

## EXERCISE 24-2   CREATING A CUSTOM WINDOW

**1.** Start Revit 2023.

**2.** From the starting page, under Families, click New.

**3.** Select Metric Window with Trim.rft (Window with trim.rft) from the English (English_I) folder.

**4.** Save As the new family in your exercise folder using the name My_Arch_Window.rfa.

**5.** Go to the Exterior elevation view.

**6.** Click the Family Types button, and set Width to be 3000mm (8'-0"); click OK.

**7.** Move your mouse over the frame and press [Tab] a couple of times until you see the tooltip **Opening Cut**; click it. Using the context tab, click the **Edit Sketch** button.

**8.** From the Draw panel, select the Start-End-Radius Arc and draw an arc as shown below. Delete the other horizontal and vertical lines. Using the Annotate tab, add a radius dimension, select it, and add a parameter called **Radius**, and equalize it to Height (for imperial the text should read Radius = 4'-0"):

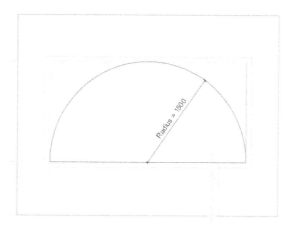

**9.** When done click (✓).

**10.** Correct the Trim : Extrusion to fit the new shape. Thickness of trim is 75mm (0'-3").

11. Look at your new window using 3D view.

12. Go to Ref. Level floor plan view.

13. Add a new reference plane at the center of the wall, and call it **Glass**.

14. Go to the Interior elevation view.

15. Go to the Create tab, locate the Work Plane panel, and click the Set button; from the drop-down list select Glass as your current plane.

16. Start the Extrusion command, using the Pick Lines tool, click the arc from inside, and lock it, then the horizontal line and lock it. From Properties set the Extrusion Start = 0, and Extrusion End = 12 (0'-1"). Set Material = Glass.

17. When done click (✓).

18. Look at your new windows using 3D view.

19. Save the new window.

20. Start a new project.

21. Draw a wall.

22. Return to the Family Editor and load the family into your project.

23. Add several windows to project. Change to Shaded so you can distinguish the glass from the other materials.

24. Close the project without saving and close the family editor.

## CREATING CUSTOM RAILINGS

- A railing family is system family, hence, you need to start with an existing family and duplicate it.
- Creating a railing requires that you understand how Revit sets the different components together to form a new railing. The components of a railing are (as shown below):
  - Post
  - Baluster Panel

- Baluster
- Rail

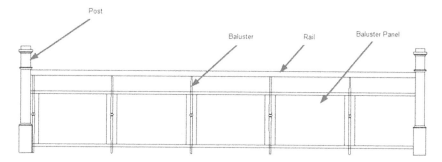

- To create a new railing type, do the following steps:
  - Load or create all components of a railing type.
  - Select one of the existing railing types, using Properties, click Edit Type button, and then click Duplicate button.
  - Under **Type Parameters** control **Rail Structure (Non-Continuous)** which control the horizontal parts of the railing, and **Baluster Placement** which controls the vertical parts of the railing.

Type Parameters

| Parameter | Value |
|---|---|
| **Construction** | ⌄ |
| Railing Height | 900.0 |
| Rail Structure (Non-Continuous) | Edit... |
| Baluster Placement | Edit... |
| Baluster Offset | 0.0 |
| Use Landing Height Adjustment | No |
| Landing Height Adjustment | 0.0 |
| Angled Joins | Add Vertical/Horizontal Segments |
| Tangent Joins | Extend Rails to Meet |
| Rail Connections | Trim |

### Rail Structure (Non-Continuous)

■ To control Rail Structure (Non-Continuous), do the following steps:
  • Click the **Edit** button next to the Rail Structure (Non-Continuous).
  • You will see the following dialog box:

  • Click **Insert** button to add a new rail to the list. To use the same rail more than one time with different height use Duplicate button. To remove one of the existing rails, click Delete. Use Up and Down buttons to rearrange the rails as desired.
  • The information needed for each row are:

| Name | Input a name of your choice |
|------|------------------------------|
| **Height** | Height for the rail measured from the base |
| **Offset** | Distance of the rail from inside or outside |
| **Profile** | The profile of the rail. Select a profile from the drop-down list. You need to load the profile before using it. |
| **Material** | Set the desired material for the rail |

### Baluster Placement

■ To control Baluster Placement, do the following steps:
  • Click the **Edit** button next to the Baluster Placement.

- You will see the following dialog box:

- Under the **Main pattern** part, add / change baluster and baluster panel families. Set Base, Base offset, Top, Top offset, Distance from previous, and Offset (to the inside or outside).
- **Set Break Pattern at**: Each Segement End, Angles Greater Than (set the angle), or Never.
- **Set the Justification criteria**: Beginning, End, Center, or Spread Pattern To Fit.
- **Set the Excess Length Fill**: None, Truncate Pattern, or set a baluster family.
- If your railing type will be used in stairs, the default number of baluster per tread is one. If you want more, specify the desired number per tread, along with baluster family.
- Under Posts part, set the family of the post for the three places, Start, Corner, and End. Set Base, Base offset, Top, Top offset, Space, and Offset (to the inside or outside).
- Set the Corner Post at whether Each Segement End, Angles Greater Than (set the angle), or Never.

## EXERCISE 24-3    CREATING CUSTOM RAILINGS

**1.** Start Revit 2023.

**2.** Open the file **Exercise 24-3.rvt**.

**3.** You will find a railing. Look at it in 3D and get acquainted with its components.

**4.** Go to **Insert** tab, locate **Load from Library** panel, and click **Load Family** button.

**5.** Go to the Profile folder, then the Railings folder, and load the following families:

  **a.** M_Elliptical Raill.rfa (Elliptical Raill.rfa)

**6.** Click Load Family again, go to the Railings folder, then the Balusters folder, and load the following families:

  **a.** M_Baluster - Custom2.rfa (Baluster - Custom2.rfa)

  **b.** M_Baluster Panel - Glass w Brackets.rfa (Baluster Panel - Glass w Brackets.rfa)

  **c.** M_Post - Newel.rfa (Post - Newel.rfa)

**7.** Select the railing, and duplicate it, naming it My_Railing1.

**8.** Click Edit next to Rail Structure (Non-Continuous).

**9.** Delete Rail 3 and Rail 4.

**10.** Change the profile of Rail 1, and Rail 2 to be M_Elliptical Raill : 40x30mm (Elliptical Rail : 1 1/2" × 1 1/8").

**11.** For Rail 1, set the Height = 650 (2'-3"), and for Rail 2, set the Height = 600 (2'-0").

**12.** Click OK and check your railing in 3D.

**13.** Select the railing again, click Edit Type, then click Edit next to Baluster Placement.

**14.** Under Main Pattern, set the Regular baluster = M_Baluster - Custom2 : 25mm (Baluster - Custom2 : 1").

**15.** Duplicate it, and change the Baluster family to M_Baluster Panel - Glass w Brackets : 600mm w 25mm Gap (Baluster Panel - Glass w Brackets : 24" w 1" Gap).

**16.** Set the distances as shown below:

Main pattern

| | Name | Baluster Family | Base | Base offset | Top | Top offset | Dist. from previous | Offset | |
|---|---|---|---|---|---|---|---|---|---|
| 1 | Pattern start | N/A | N/A | N/A | N/A | N/A | N/A | N/A | Delete |
| 2 | Regular balust | M_Baluster - Custom2 : 25m | Host | 0.0 | Top Rail Elem | 0.0 | 310.0 | 0.0 | Duplicate |
| 3 | Regular balust | M_Baluster Panel - Glass w Br | Host | 0.0 | Rail 2 | -50.0 | 310.0 | 0.0 | Up |
| 4 | Pattern end | N/A | N/A | N/A | N/A | N/A | 0.0 | N/A | Down |

Main pattern

| | Name | Baluster Family | Base | Base offset | Top | Top offset | Dist. from previous | Offset | |
|---|---|---|---|---|---|---|---|---|---|
| 1 | Pattern star | N/A | N/A | N/A | N/A | N/A | N/A | N/A | Delete |
| 2 | Regular bal | Baluster - Custom2 : 1" | Host | 0' 0" | Top Rail El | 0' 0" | 1' 0 51/128 | 0' 0" | Duplicate |
| 3 | Regular bal | Baluster Panel - Glass | Host | 0' 0" | Rail 2 | -0' 2" | 1' 0 51/128 | 0' 0" | Up |
| 4 | Pattern en | N/A | N/A | N/A | N/A | N/A | 0' 0" | N/A | Down |

**17.** Set Justify = Spread Pattern To Fit.

**18.** Under Posts set Start Post and End Post = M_Post - Newel : 150mm (Post - Newel : 6 1/4").

**19.** Click OK, twice, and look at your railing in 3D.

**20.** Go to the South elevation view, and set Visual Style = Shaded, and look at the railings.

**21.** Save and close the file.

# NOTES

## CHAPTER REVIEW

1. You can create a window from scratch using an RFA template file:

   **a.** True

   **b.** False

2. _____ allows the user to control when elements of your family will be shown, and when will be hidden.

3. While you are defining a custom railing family, you can set the rail height from base or top:

   **a.** True

   **b.** False, only from top

   **c.** False, only from base

   **d.** True, but only for certain profiles only

4. Door families are good enough to cover all of your needs; no need to customize door families:

   **a.** True

   **b.** False

5. While defining custom railing families, all of the following is true, except one statement:

   **a.** You can define the material of the rails

   **b.** You can define the material of the balusters

   **c.** You cannot define the material of the balusters

   **d.** A baluster has a family RFA file which contains the material

6. In Railing definition, control, _____, Baluster Panel, Baluster, and Rail.

## CHAPTER REVIEW ANSWERS

1. b

3. c

5. b

## CHAPTER 25

# WORKSETS AND SHARED VIEWS

## This Chapter Contains

- Creating and controlling worksets in Revit
- Creating and viewing Shared Views

## INTRODUCTION TO WORKSETS

- Since the Revit model is in a single file, and since no single person can finish the designing process by themself, we need to share the project with other colleagues.
- The idea is very simple:
  - There will be a single Central File that resides in the server.
  - Each user will create a copy from it in their own computer; we will call this file the Local file.
  - You will make the changes and save it back to the Central File.

- BIM Manager can create ownership for each element in the project. If other people want to make changes to your owned elements, they have to receive your permission.

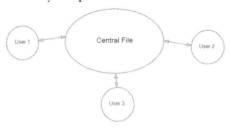

## HOW TO CREATE A CENTRAL FILE

- A Central file is a normal RVT file
- When you start **Collaborate** command, it will become a Central File.
- Assume we opened a normal RVT file that contains three levels, do the following steps:
  - Go to **Collaborate** tab, locate **Manage Collaboration** panel, and select **Collaborate** button:

- You will see the following dialog box:

- This dialog box explains the collaboration concept: "This will allow multiple people to work on the same Revit model simultaneously."
- It gives you two options to select from: **Collaborate within your network** or **In the cloud**.
- Select the first choice and click OK.
- In the **Manage Collaboration** panel, click the **Worksets** button, and you will see the following dialog box. Revit tells you it created two worksets automatically, one for levels and grids, and the second for all other elements called Workset1:

- Using the three buttons at the top right, you can create a **New** workset, or **Delete** or **Rename** an existing workset.
- When you click **New** button, you will receive a dialog box as shown here:

- Type the name of the new workset and click OK.

### How to Assign Elements to Worksets

- Creating worksets on its own means nothing. You need to assign a workset for each group of elements. Do the following steps:
  - Go to a suitable view.
  - Select the desired elements.

- Using Properties, set the name of the workset:

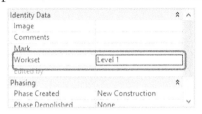

- Repeat this process for as many elements as you wish.
- As the creator of the Central file, all sets are editable by you, and owned by you; look at the following example:

- The creator of the central file needs to release the ownerships of all worksets, or other users will not be able to work at their local files.
- To do that go to **Collaborate** tab, locate **Synchronize** panel, and click **Relinquish All Mine** button (the file should be saved first):

## CREATING A LOCAL FILE

- A local file is a copy of the central file opened in the user computer.
- You can add to and modify the existing elements. Once you are done, save back to the central file to update it.
- You can create a local file each time you start Revit or you can save the local file to be used frequently.
- To create a local file, do the following steps:
  - Start the Open command and select the central file.

- At the bottom of the dialog box, locate **Create New Local** checkbox, and turn it on, as shown here:

- Click **Open** button.
- A copy of the project is created in your local computer. It will hold the same name as the central file with your Revit Username added as a suffix.
- Use the Save As command to give the file a new name.
- Go to **Collaborate** tab, locate **Manage Collaboration** panel, and click **Worksets** button, and you will see the following:

| Name | Editable | Owner | Borrowers | Opened | Visible in all views |
|---|---|---|---|---|---|
| Level 1 | No | | | Yes | ☑ |
| Level 2 | No | | | Yes | ☑ |
| Level 3 | No | | | Yes | ☑ |
| Shared Levels and Grids | No | | | Yes | ☑ |
| Workset1 | No | | | Yes | ☑ |

- There are no Owners for any workset; all you have to do is to make the workset Editable and assign your name to be the workset owner:

| Name | Editable | Owner | Borrowers | Opened | Visible in all views |
|---|---|---|---|---|---|
| Level 1 | No | | | Yes | ☑ |
| Level 2 | Yes | MM2 | | Yes | ☑ |
| Level 3 | No | | | Yes | ☑ |
| Shared Levels and Grids | No | | | Yes | ☑ |
| Workset1 | No | | | Yes | ☑ |

- If you keep your ownership to this workset, nobody can modify it except after taking your permission.

## FUNCTIONS TO HELP CONTROL WORKSETS

- There are lots of functions and tools to help you control the worksets process. Each one of them is unique by itself. They are:

### Active Workset

- This list will tell you which workset is available for editing.
- You can reach it from two different places:

- Go to **Collaborate** tab, locate **Manage Collaboration** panel, and select the desired workset. Revit will tell which is available for editing:

- The second way is to use the bar at the bottom of the screen, where you will find the same list:

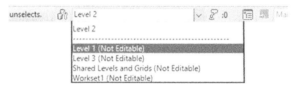

### Gray Inactive Worksets

- Based on the active workset, Revit can gray out the other inactive worksets, to show them in dimmed color.
- Go to **Collaborate** tab, locate **Manage Collaboration** panel, and select **Gray Inactive Worksets** button:

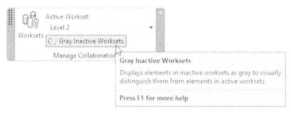

- You will receive a result similar to the following:

### Synchronize

- Synchronizing means updating the central file with your local file changes.
- There are two types:
  - Synchronize and Modify Settings
  - Synchronize Now
- Go to **Collaborate** tab, locate **Synchronize** panel, and select one of the two options:

- You can reach the same two options using Quick Access Toolbar, at the top left part of the screen:

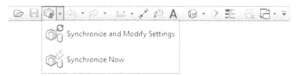

### Reload Latest

- It loads changes from the central file to your local file.
- Go to **Collaborate** tab, locate **Synchronize** panel, and select **Reload Latest** button:

## Worksharing Display Options

- Using the Status bar, you can activate the Worksharing Display options:

- You can show colors for Checkout status, owners, model updates, and worksets. You can turn off the worksharing display or control the display settings.
  - To control the display settings, click Worksharing Display Settings, and you will see the following dialog box:

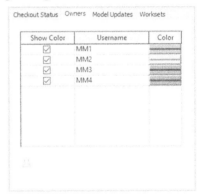

  - You can show these colors on your screen, using any of the methods mentioned previously.

## EXERCISE 25-1    CREATING AND CONTROLLING WORKSETS

**1.** Start Revit 2023.

NOTE *If you cannot change your username, you will not be able to do this exercise.*

**2.** Go to File menu, select **Options**, under **General**, change the Username to be Yourname_Manager (we will change the username several times through this exercise, so if the username is important to you, make a note, at the end of the exercise, change it back to the original name).

**3.** Open the file **Exercise 25-1.rvt**.

**4.** Save as the project **Yourname_Central.rvt**.

**5.** Go to **Collaborate** tab, locate **Manage Collaboration** panel, click **Collaborate** button, select the first option, and click OK.

**6.** Start Worksets command. Look at the created worksets. Create three new worksets, and call them Level 1, Level 2, and Level 3 (take a note of the name of the owner of the old and new worksets). When done, click OK.

**7.** Go to the North elevation view. Change the Visual Style to Wireframe.

**8.** Select everything in Level 1 (excluding the level) including the floor. In Properties change the Workset to Level 1.

**9.** Select everything in Level 2 (excluding the level) including the floor. In Properties change the Workset to Level 2.

**10.** Select everything in Level 3 (excluding the level) including the floor. In Properties change the Workset to Level 3.

**11.** Go to **Collaborate** tab, locate **Manage Collaboration** panel, and click **Gray Inactive Workset** button, from the list select Level 1. See how all of the other levels gray out except Level 1. Try it for Level 2 and Level 3. Then turn off **Gray Inactive Workset** button.

**12.** Save the file and read the message from Revit which tells you that your file now becomes the central file.

**13.** Go to **Collaborate** tab, locate **Synchronize** panel, and click **Relinquish All Mine** button to release the ownership of all worksets (if you do not do that, other users cannot work with worksets).

**14.** Close the file.

**15.** Using Options, change the username to be MH1.

**16.** Using File menu, select Open, then select yourname_Cenral.rvt (use this method) but do not click OK; make sure that Create New Local checkbox is turned on, then click Open.

**17.** Check the name of the file (it should look like yourname_Central_MH1. rvt).

**18.** Go to Worksets command, and make Level 1 Editable (you will see that the owner will be MH1).

**19.** Save the file and Close it. Once Revit asks you if you want Relinquish or keep your ownership, choose Keep your ownership.

**20.** Repeat this process as follows:

    **a.** Username = MH2, owns Level 2

    **b.** Username = MH3, owns Level 3

**21.** In the third file, you should receive the following:

| Name | Editable | Owner | Borrowers | Opened | Visible in all vi |
|---|---|---|---|---|---|
| Level 1 | No | MH1 | | Yes | ☑ |
| Level 2 | No | MH2 | | Yes | ☑ |
| Level 3 | Yes | MH3 | | Yes | ☑ |
| Shared Levels and Grids | No | | | Yes | ☑ |
| Workset1 | No | | | Yes | ☑ |

**22.** Save and close the MH3 file.

**23.** Change the username to MH2.

**24.** Open the file yourname_Central_MH2.rvt.

**25.** Go to the 3D view and rotate the project until you see the outside door in Level 1.

**26.** Select it and move it to the right by 2700 mm (9'-0").

**27.** Did Revit accept the move and why? _____

**28.** Do not send a request, click Cancel.

**29.** Close it without saving but keep the ownership.

**30.** Change the username to be yourname_Manager, and open the central file (make sure you are not opening it as a local file).

**31.** Go to the South elevation view and change the height of Level 2 to be 3800mm (13'-0").

**32.** Close the file and answer Yes to save the file.

**33.** Change the username to be MH2, and open yourname_Central_MH2.rvt.

**34.** Go to South elevation view; what is the height of Level 2? _____

**35.** Go to **Collaborate** tab, locate **Synchronize** panel, and click **Reload Latest** button.

**36.** What is the height of Level 2 after this action? _____

**37.** Go to 3D view.

**38.** Click on Worksharing Display button and select Owners, and then select Worksets.

**39.** Select Worksharing Display Settings and change the colors as you wish.

**40.** Save and close keeping your ownership.

## SHARED VIEWS—INTRODUCTION

- Sharing your design by exchanging RVT files with other persons or firms, is very risky. Yet, in today's business environment, you are obliged to send some data to many stakeholders in order for them to view, approve, and tell you the next step.
- Revit offers you the best solution for this dilemma! You can send stakeholders 2D & 3D views of your RVT without sending them the actual one. Using Shared Views in Revit will enable you to do the following:
  - Not risking your design by sending the original RVT file
  - No need for the recipients to have Revit installed in their machines
  - No need to start new accounts, no usernames, no passwords
  - You need only a browser to view the Shared 2D & 3D Views
- You will use the Autodesk Viewer which is completely free-of-charge, yet, you can use it without signing in or creating a new account.

## SHARED VIEWS

- ■ To create a shared view, do the following steps:
  - • Open your desired file and go to the desired 2D or 3D view
  - • Go to **Collaborate** tab, locate **Share** panel, and click **Shared Views** button:

- • You will see the Shared Views palette, similar to the following:

- Go to the desired view, click New Shared View button, you will see the following dialog box:

- Type in the name of the view, and click Share button to save the view to the cloud. You will see the following message at the left right side of your screen:

- The above message will disappear once the uploading is complete. The following message will appear at the same place:

- If you want to view the uploaded view in the browser, click the first option. If you want to copy the link to send it to your stakeholders, click the second option.

- In the browser, you will see the following:

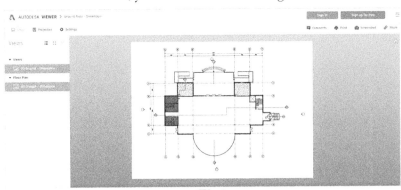

- From the top right, click to sign in (if you have an account), or Sign up for free (if you don't have an account)
- At the top left you can switch on Views list (two methods of display), Properties, and Settings
- At the bottom of the screen, click Measure, you will see an additional menu similar to the following:

- It will allow you to measure lots of things in the view, once done click Done button
- You can use Zoom and Pan (you can use the mouse as well) to navigate through the view. Also, use Fit to see the whole view, and Home to see the home view
- At the top right, click Comment (for you to post comments you need to be Signed In) the Markup tool will appear to the right of the Measure tool, once you click it, you will see the following menu:

- Select the color and the weight then select the tool you want to use to comment, choose from Pencil, Arrow, Cloud, or Text. You can use the Undo and Redo if you prefer

**NOTE** *You can print the view and you can take a screen shot of it.*

**Follow Up on Comments**

- When one of the stakeholders comment on one of the shared views, the owner of shared view will be notified as shown below. You can reply to the comment and the stakeholder will see it:

- At the bottom of the palette click Reply and type in your reply to the stakeholder comment

## EXERCISE 25-2   SHARED VIEWS

**1.** Start Revit 2023.

**2.** Open the file **Exercise 25-2.rvt**.

**3.** Go to 00 Ground – Dimension floor view

**4.** Start Shared Views command

**5.** From Shared Views palette click New Shared View button, and call it Ground Floor – Dimension, then click Share button

**6.** Wait until the uploading is finished

**7.** Once the message appears click View in the Browser option

8. It will take you to your default browser and show you the view you just uploaded.

9. Test the zooming and the panning

10. Measure the area of the right Managers office

11. Measure the length and the width of one of the toilets

12. (If you want to comment, you should Sign In first) zoom in to the upper toilet, and click Markup, change the Weight to the smallest. Change the color to Magenta, then create a cloud over the upper toilet

13. Using the Text tool, type the following text: "Please check the dimensions as they don't match the original design" click anywhere out. At the top right click Save button

14. Go back to Revit, using the Shared Views palette, at the top right click the two circular arrows to refresh, you will discover there is a comment. You can reply to it by saying: "I will check and get back to you"

15. Save the file and close both the Revit file and the browser

# NOTES

# CHAPTER REVIEW

**1.** To gray out the inactive worksets, to show them in dimmed color, use Gray Inactive Workset command:

    **a.** True

    **b.** False.

**2.** Using _____ you can activate the Worksharing Display options.

**3.** One of the following is not among the Worksets commands:

    **a.** Synchronize

    **b.** Relinquish all not mine

    **c.** Reload latest

    **d.** Gray Inactive worksets

**4.** For each owner of a Workset, you can assign a color:

    **a.** True

    **b.** False.

**5.** To create a local file from Central File:

    **a.** Go to Application Menu, select Open, then select Open Local.

    **b.** Type the OPENLOCAL command.

    **c.** In the Open dialog box, turn on Create New Local checkbox.

    **d.** In the New dialog box, turn on Create New Local checkbox.

**6.** You can show colors for Checkout status, owners, model updates, and worksets, using _____.

**7.** You can share only 2D views:

    **a.** True

    **b.** False.

**8.** In order to comment on a shared folder, you should Sign In first:

    **a.** True

    **b.** False.

## CHAPTER REVIEW ANSWERS

**1.** a

**3.** b

**5.** c

**7.** b

# INDEX